Macmillan/McGraw-Hill S

STRUCTURES OF LIFE

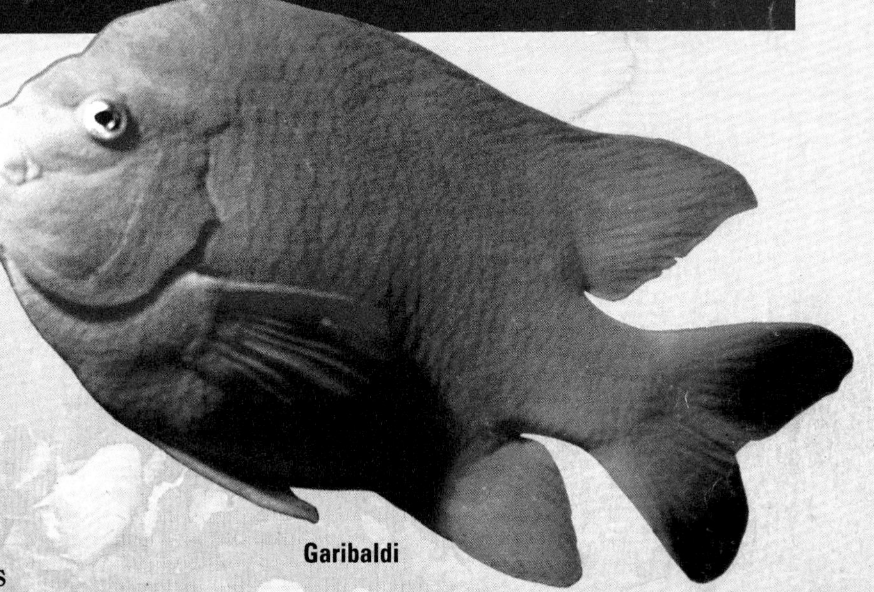

Garibaldi

AUTHORS

Mary Atwater
The University of Georgia
Prentice Baptiste
University of Houston
Lucy Daniel
Rutherford County Schools
Jay Hackett
University of Northern Colorado
Richard Moyer
University of Michigan, Dearborn
Carol Takemoto
Los Angeles Unified School District
Nancy Wilson
Sacramento Unified School District

Macmillan/McGraw-Hill School Publishing Company
New York Chicago Columbus

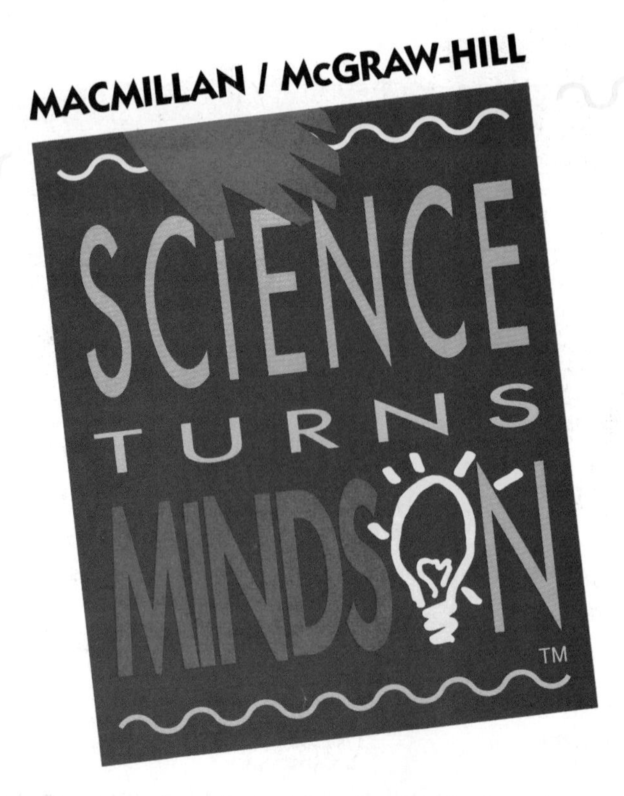

CONSULTANTS

Assessment:

Janice M. Camplin
Curriculum Coordinator, Elementary Science
Mentor, Western New York
Lake Shore Central Schools
Angola, NY

Mary Hamm
Associate Professor
Department of Elementary Education
San Francisco State University
San Francisco, CA

Cognitive Development:

Dr. Elisabeth Charron
Assistant Professor of Science Education
Montana State University
Bozeman, MT

Sue Teele
Director of Education Extension
University of California, Riverside
Riverside, CA

Cooperative Learning:

Harold Pratt
Executive Director of Curriculum
Jefferson County Public Schools
Golden, CO

Earth Science:

Thomas A. Davies
Research Scientist
The University of Texas
Austin, TX

David G. Futch
Associate Professor of Biology
San Diego State University
San Diego, CA

Dr. Shadia Rifai Habbal
Harvard-Smithsonian Center for Astrophysics
Cambridge, MA

Tom Murphree, Ph.D.
Global Systems Studies
Monterey, CA

Suzanne O'Connell
Assistant Professor
Wesleyan University
Middletown, CT

Environmental Education:

Cheryl Charles, Ph.D.
Executive Director
Project Wild
Boulder, CO

Gifted:

Sandra N. Kaplan
Associate Director, National/State Leadership
Training Institute on the Gifted/Talented
Ventura County Superintendent of Schools Office
Northridge, CA

Global Education:

M. Eugene Gilliom
Professor of Social Studies and Global Education
The Ohio State University
Columbus, OH

Merry M. Merryfield
Assistant Professor of Social Studies and Global Education
The Ohio State University
Columbus, OH

Intermediate Specialist

Sharon L. Strating
Missouri State Teacher of the Year
Northwest Missouri State University
Marysville, MO

Life Science:

Carl D. Barrentine
Associate Professor of Biology
California State University
Bakersfield, CA

V.L. Holland
Professor and Chair, Biological Sciences Department
California Polytechnic State University
San Luis Obispo, CA

Donald C. Lisowy
Education Specialist
New York, NY

Dan B. Walker
Associate Dean for Science Education and Professor of Biology
San Jose State University
San Jose, CA

Literature:

Dr. Donna E. Norton
Texas A&M University
College Station, TX

Tina Thoburn, Ed.D.
President
Thoburn Educational Enterprises, Inc.
Ligonier, PA

Macmillan/McGraw-Hill School Division
10 Union Square East
New York, New York 10003

Printed in the United States of America

ISBN 0-02-274274-3 / 5

3 4 5 6 7 8 9 VHJ 99 98 97 96 95 94 93

Moose

Mathematics:

Martin L. Johnson
Professor, Mathematics Education
University of Maryland at College Park
College Park, MD

Physical Science:

Max Diem, Ph.D.
Professor of Chemistry
City University of New York, Hunter College
New York, NY

Gretchen M. Gillis
Geologist
Maxus Exploration Company
Dallas, TX

Wendell H. Potter
Associate Professor of Physics
Department of Physics
University of California, Davis
Davis, CA

Claudia K. Viehland
Educational Consultant, Chemist
Sigma Chemical Company
St. Louis, MO

Reading:

Jean Wallace Gillet
Reading Teacher
Charlottesville Public Schools
Charlottesville, VA

Charles Temple, Ph.D.
Associate Professor of Education
Hobart and William Smith Colleges
Geneva, NY

Safety:

Janice Sutkus
Program Manager: Education
National Safety Council
Chicago, IL

Science Technology and Society (STS):

William C. Kyle, Jr.
Director, School Mathematics and Science Center
Purdue University
West Lafayette, IN

Social Studies:

Mary A. McFarland
Instructional Coordinator of Social Studies, K-12, and Director of Staff Development
Parkway School District
St. Louis, MO

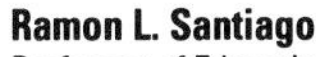

Students Acquiring English:

Mrs. Bronwyn G. Frederick, M.A.
Bilingual Teacher
Pomona Unified School District
Pomona, CA

Misconceptions:

Dr. Charles W. Anderson
Michigan State University
East Lansing, MI

Dr. Edward L. Smith
Michigan State University
East Lansing, MI

Multicultural:

Bernard L. Charles
Senior Vice President
Quality Education for Minorities Network
Washington, DC

Cheryl Willis Hudson
Graphic Designer and Publishing Consultant
Part Owner and Publisher, Just Us Books, Inc.
Orange, NJ

Paul B. Janeczko
Poet
Hebron, MA

James R. Murphy
Math Teacher
La Guardia High School
New York, NY

Ramon L. Santiago
Professor of Education and Director of ESL
Lehman College, City University of New York
Bronx, NY

Clifford E. Trafzer
Professor and Chair, Ethnic Studies
University of California, Riverside
Riverside, CA

STUDENT ACTIVITY TESTERS

Jennifer Kildow
Brooke Straub
Cassie Zistl
Betsy McKeown
Seth McLaughlin
Max Berry
Wayne Henderson

FIELD TEST TEACHERS

Sharon Ervin
San Pablo Elementary School
Jacksonville, FL

Michelle Gallaway
Indianapolis Public School #44
Indianapolis, IN

Kathryn Gallman
#7 School
Rochester, NY

Karla McBride
#44 School
Rochester, NY

Diane Pease
Leopold Elementary
Madison, WI

Kathy Perez
Martin Luther King Elementary
Jacksonville, FL

Ralph Stamler
Thoreau School
Madison, WI

Joanne Stern
Hilltop Elementary School
Glen Burnie, MD

Janet Young
Indianapolis Public School #90
Indianapolis, IN

CONTRIBUTING WRITER

Rosalyn Vu

STRUCTURES OF LIFE

Lessons **Themes**

Activities!

EXPLORE

TRY THIS

Features

Links

GLOBAL VIEW

CAREERS

SCIENCE TECHNOLOGY AND Society

Departments

STRUCTURES OF LIFE

How many living things do you see in this picture? There are some living things, smaller than the smallest flea, you won't be able to see. But we know they are there. People have estimated that today there are more than 10 million different kinds of living things on Earth. We have only identified 1.5 million of them. And there are many more organisms that have become extinct than are living on or in Earth's land, water, or air today.

What do you and all other living things, past or present, have in common? What is life made of? These are the questions you'll explore in this unit.

Moose, Denali National Park, Alaska

Earth is over 4.6 billion years old. The first life on Earth appeared about 3.5 billion years ago. Over many years, some types of organisms that have existed on Earth have become extinct. Could you have something in common with the first life on Earth?

In this unit you'll examine what makes up living things and how their parts and systems function to live. You'll also discover a world of living things too small to be seen with your eyes. You'll also look at how scientists try to group all of the living things by their structures of life.

Parts of Life

What makes up living things?

What You Need

pencil, *Activity Log* page 1

The next time you go outside, look for two different living things to examine. Draw a picture of each organism in your ***Activity Log***. Then label all the parts you observe. Look at one part and explain what you think makes up the part. Discuss your explanation with your classmates.

Microscopic organisms

Science in Literature

All living things struggle to survive. These books explore how different kinds of organisms live.

Biology: Plants, Animals, and Ecology **by Ifor Evans.**
New York: Watts, 1984.

Living things are everywhere. This book is all about biology, the science of living things. Even though living things seem different, they have much in common. Use this colorful book as you study this unit to find out more about the structures of life.

A Time to Fly Free
by Stephanie Tolan.
New York: Macmillan, 1983.

"The great thing about animals . . . is that they never seem to waste energy worrying. . . . I wish I could be like that," said ten-year-old Josh Taylor. Josh is the main character in the book *A Time to Fly Free.* He is fascinated with birds and other wildlife around him. Read this book to find out what Josh learns about survival from the living organisms around him. In your ***Activity Log*** on page 2 explain what you think is a great thing about animals.

Other Good Books To Read

A Golden Guide of Birds
by Herbert Zim and Ira Gabrielson.
New York: Golden Press, 1956.

A Golden Guide of Birds pictures 129 of the most familiar American birds. It is one of many books in the series that are available to help identify plants, animals, and other living things. Each book has descriptions of the specimens and maps showing where they can be found. Take one of these helpful guides with you the next time you set out on a field trip.

Cry of the Crow
by Jean Craighead George.
New York: Harper, 1980.

Cry of the Crow is an interesting story about a young girl who finds a baby crow. She faces a hard decision when she realizes she can't keep the crow as a pet. Read this book to see what happens to both of them.

The Human Body
by Mary Elting.
New York: Aladdin Books, 1986.

The human body is complex. All of its parts function together to make your 100 trillion cells do what they do. Explore the structure of the human body by reading this book.

Why Does My Nose Run?
by Nancy Baggett and Joanne Settel.
New York: Atheneum, 1985.

Why does your nose run? Why do you sneeze? Why do you yawn? What causes warts, moles, and pimples? What are hiccups? This book answers these questions and many more.

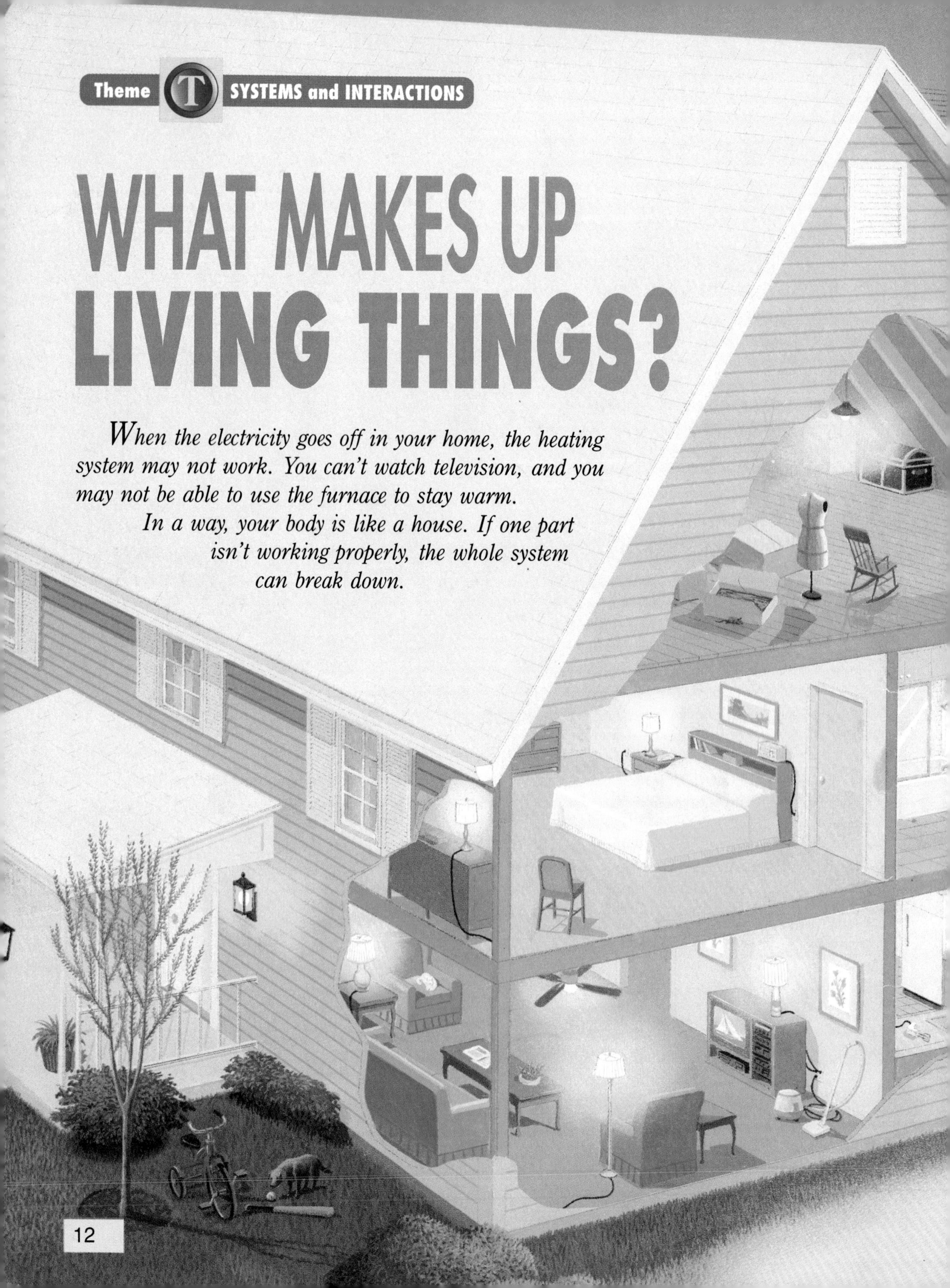

Theme T SYSTEMS and INTERACTIONS

WHAT MAKES UP LIVING THINGS?

When the electricity goes off in your home, the heating system may not work. You can't watch television, and you may not be able to use the furnace to stay warm.

In a way, your body is like a house. If one part isn't working properly, the whole system can break down.

Electrical system of a house

You've heard the expression "a chain is as strong as its weakest link." Just like a chain, different links or systems in your home have to work together to make it possible for you and your family to live comfortably. The plumbing, heating and cooling, and electrical systems have to be in working order. Each of these systems, however, depends on smaller parts to make it work. If a mouse chews through one small wire in a furnace, the heating system can stop working. Because of that pipes could freeze, burst, and stop the plumbing system from working. If water from the broken pipes gets into the electrical system, watch out!

Living things also have systems that work together so that life processes can be carried out. If all the systems aren't working well together, the organism may not be able to survive.

Minds On! Cross your arms over your chest. Think about all the different parts of your body that worked together to do that. In your ***Activity Log*** on page 3, make a list of all the parts involved in crossing your arms. Did you remember to include your eyes and brain? Compare your list with one other person. Were your lists similar? ●

EXPLORE

Activity!

What Makes Up an Onion Bulb?

How many parts does an onion plant have? What are they made of?

What You Need

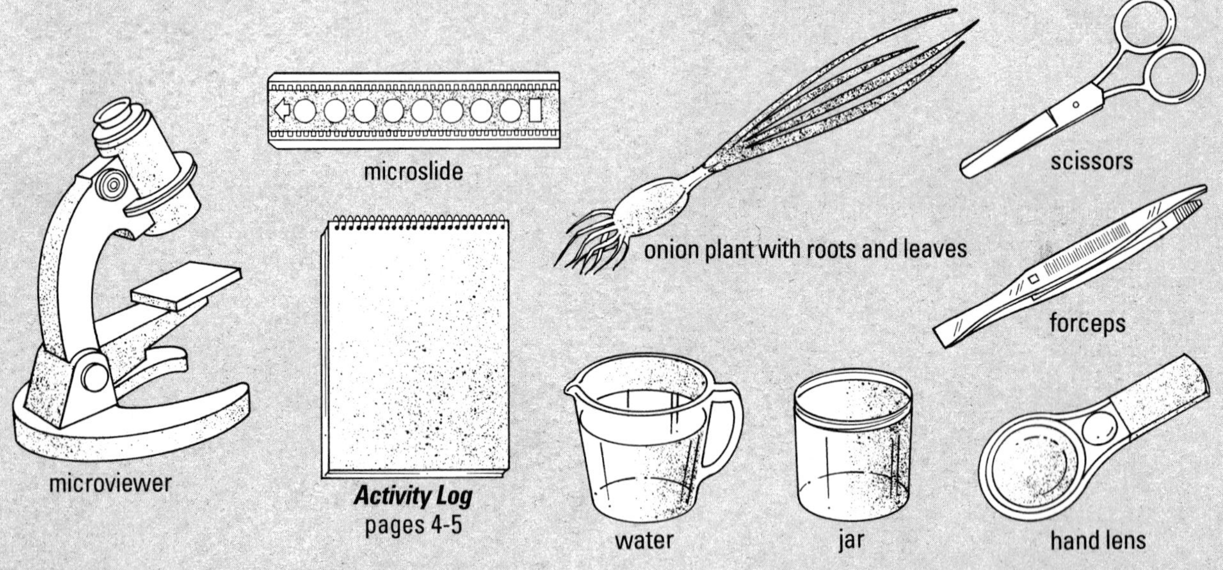

What To Do

1 Look at the whole onion plant carefully and sketch its parts in your ***Activity Log***.

2 Have your teacher cut the onion plant in half lengthwise. Again draw what you see.

3 Use scissors to cut off one of the roots. Examine the root with the hand lens. Look carefully at the cut end. What can you see? Draw it in your ***Activity Log***.

4 Place the bottom part of the onion plant in cold water so that its layers are easy to separate. Cut a 5-mm section of a layer with your scissors.

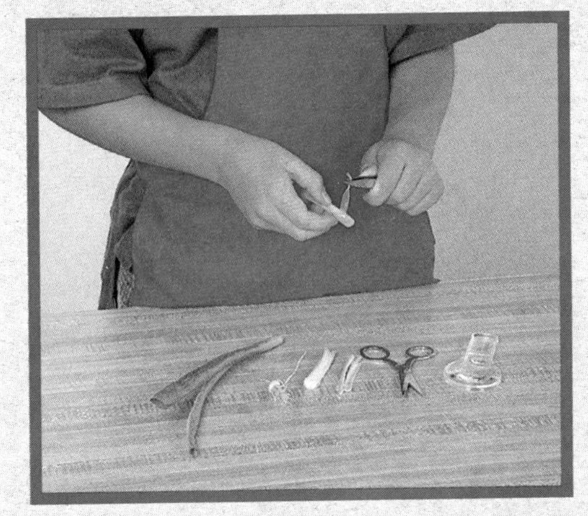

5 Remove a thin layer on the inside of the section with your forceps. Observe the layer of skin with a hand lens.

6 Look at the onion skin on a microslide with a microviewer. Sketch what you see in your ***Activity Log***. How would you describe what you see? Do the layers seem to have anything in them? Are these smaller parts about the same shape and size? Write a brief description of what you see.

7 Remove a thin piece of the green onion leaf and observe it with a hand lens.

What Happened?

1. Describe the parts of the onion plant.
2. What did the parts of the plant you saw under the microviewer look like? If you looked at a part of the root under the microviewer, would you have seen similar structures?

What Now?

1. What do the parts of an onion plant seem to have in common?
2. What parts of the onion plant make up the organism?

Organ Systems

In the Explore Activity you just did, you looked at a living thing—an onion plant. Another name for a living thing is an organism.

An organism has parts that work together. You looked at the parts of the onion—roots, stem, and leaves. You'll be asked to look back at your notes and think about the Explore Activity as you study this lesson.

Organisms are composed of many different parts that work together in systems. Think about crossing your arms. What body parts did you use? Did you think about reading the words on the page? The message went to your brain, your brain sent a message to your arm muscles, and you lifted and crossed them. Many of these actions involved the parts of your body called the nervous system.

The parts of the nervous system are the brain, spinal cord, and nerves. The nervous system works to sense what's happening around the body and to decide what to do about it. Your senses of sight, hearing, touching, smelling, and tasting are important in the proper working of the nervous system to keep you healthy and safe.

ONION PLANT SYSTEM

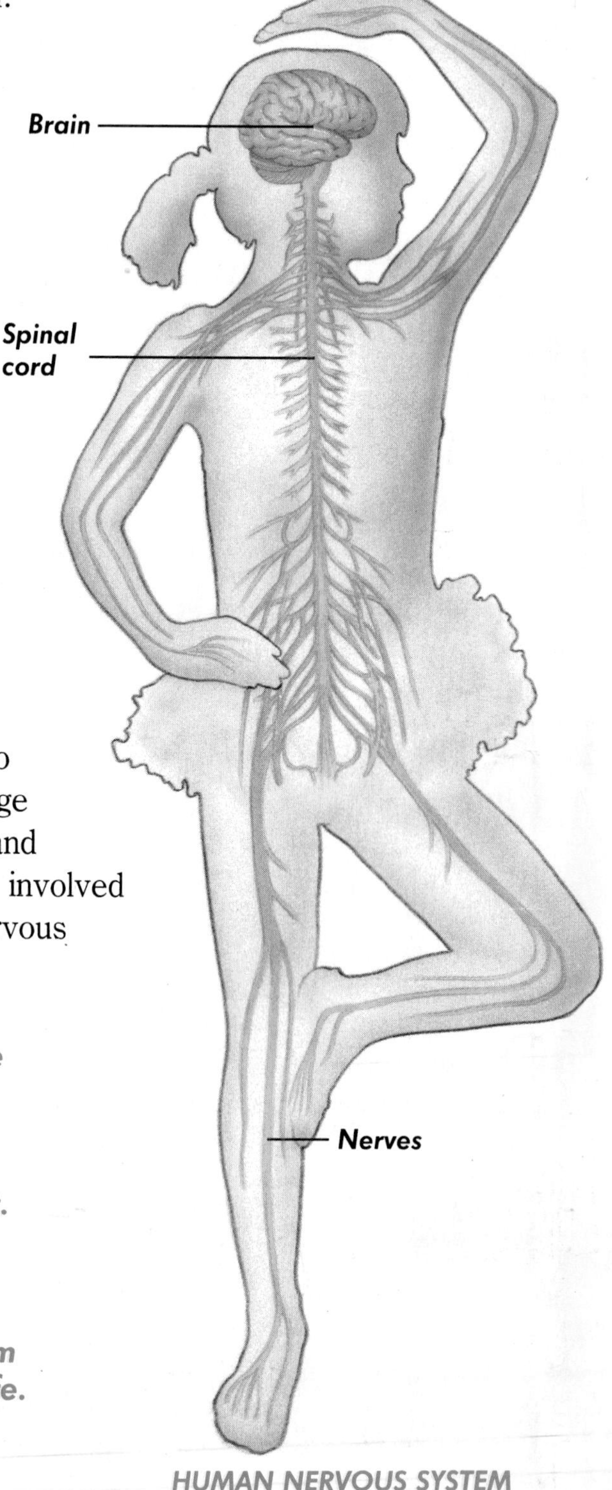

HUMAN NERVOUS SYSTEM

Minds On!

Look at this picture of many living things or organisms. How many can you find? In your ***Activity Log*** on page 6, make a list of all the organisms you see in the picture. How are they different and alike? ●

Before we talk about any more animal systems, do the following activity to look at part of a chicken.

TRY THIS **Activity!**

Systems in a Chicken Wing

What are the parts that make up systems in a chicken wing?

What You Need

cooked chicken wing, waxed paper, hand lens, plastic knife, forceps, *Activity Log* page 7

Put the chicken wing on waxed paper. Carefully separate the skin, meat, fibers, and bones with the knife. Examine with the forceps how they're connected to each other. In your ***Activity Log***, draw the parts and label them. How are the parts connected? Use the hand lens to examine each of the parts more closely. Turn the skin over and examine the underside. How do the parts differ? How do the parts work together? What can you see with the hand lens? Draw what you see.

Another system that was used in crossing your arms was the muscular system. Nerves are attached to the muscles in your arms. When the nerves in the muscles received the message to move, you raised and crossed your arms. The muscles in your body are attached to the bones of the skeletal system. Without bones, your body wouldn't have much shape and moving around would become difficult, if not impossible. Look back at the chicken-wing activity. The meat that we eat is part of the muscle of the chicken.

All the muscles together make up the muscular system.

Muscles are attached to the bones, part of the skeletal system.

You may also have found some blood vessels in the chicken wing. They are part of the circulatory system. The nervous, muscular, skeletal, and circulatory systems work together so that the chicken can live and carry out its life functions.

Do the systems in your body work alone? What would it be like if your nerves weren't attached to your muscles? Bones and muscles must connect and interact to function. Just as the electrical and plumbing systems work together in a home, the interaction of body systems makes life possible.

It's easy to look at the systems in animals and see how they work and interact. It's harder to think about plant systems and see their interactions. Look back at the Explore Activity you did. You noticed that the onion had many leaves.

The root systems of plants function with the stem and the leaves so that the life processes of the plant are carried out.
Leaf system of honey locust tree
Root system of dandelion
Leaf system of papyrus plant
Root system of corn plant
Leaves work together as a system to make food. However, the leaves can't do their job if the roots and stem aren't working together to bring water and nutrients to the leaves.
The entire fern plant couldn't live without the interactions of the systems.

Organs, Tissues, Cells

Now that you've looked at organ systems, you are ready to think about the individual parts of systems. Systems in organisms are composed of parts. These parts of systems are called **organs**.

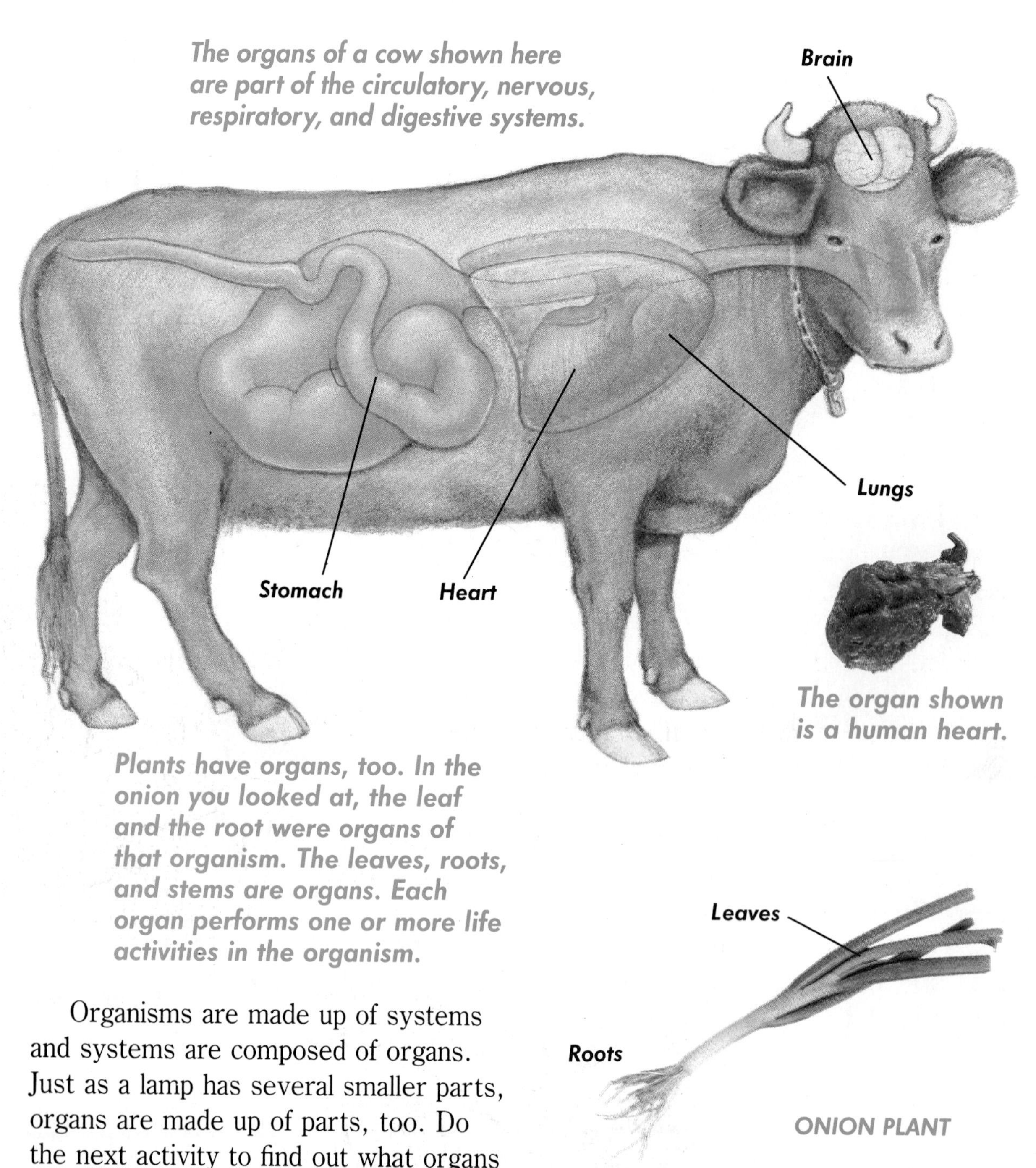

The organs of a cow shown here are part of the circulatory, nervous, respiratory, and digestive systems.

The organ shown is a human heart.

Plants have organs, too. In the onion you looked at, the leaf and the root were organs of that organism. The leaves, roots, and stems are organs. Each organ performs one or more life activities in the organism.

Organisms are made up of systems and systems are composed of organs. Just as a lamp has several smaller parts, organs are made up of parts, too. Do the next activity to find out what organs are made of.

TRY THIS Activity!

Where's the Tissue?

Find the tissues in ground beef.

What You Need

small amount of ground beef, hand lens, forceps, *Activity Log* page 8

Separate the beef as thinly as possible. Using the forceps, pick up a very small portion of the beef. Do you see tiny fibers? If not, try again. The ground beef is the muscle tissue.

Examine the fibers more closely with the hand lens. What do you see? Draw what you see in your ***Activity Log***. If the fibers are tissues what would all the fibers together be?

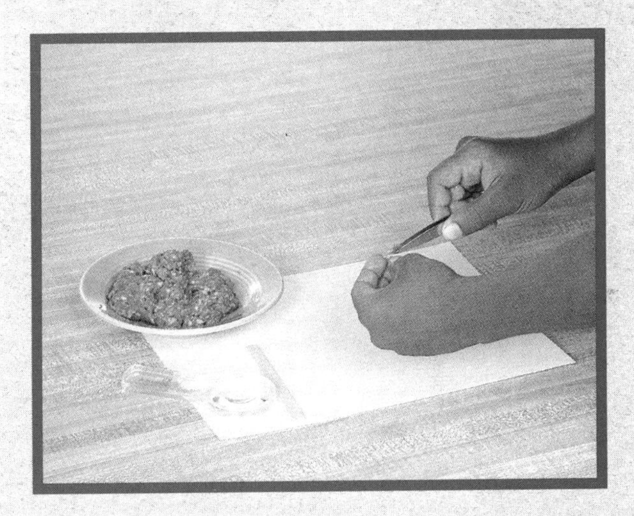

How did the ground beef in the last activity look under the hand lens? How would you describe what you saw? Different kinds of tissues in an animal's body include muscle, bone, skin, nerve, and blood. Think about the chicken wing you cut up. When you looked at the chicken muscle (meat) with your hand lens, the stringy fibers you may have noticed are tissues. Different **tissues** work together to carry out the functions of the muscles.

Look back at the notes you took when you looked at the onion plant. You observed a thin layer of onion with a hand lens. That layer was a part of the leaf tissue. If you look closely at a leaf, you may see something that looks like tracks. These are tissues that carry water with nutrients through the leaf, and also transport food that is made in the leaves.

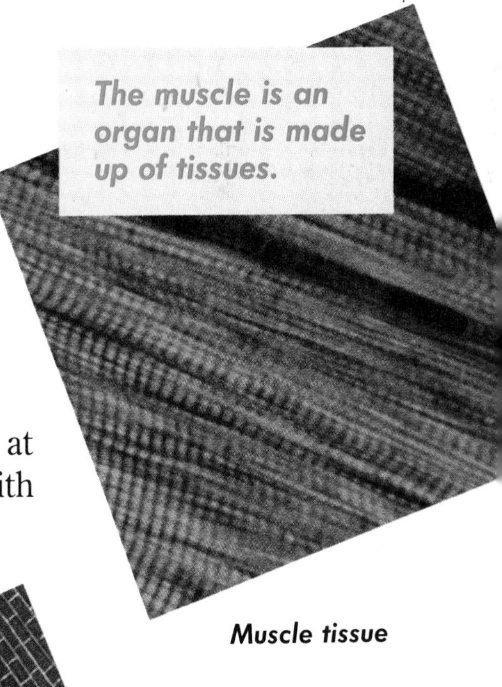

The muscle is an organ that is made up of tissues.

Muscle tissue

The tissue of a leaf is where the pores of the plant are located.

Onion leaf tissue

A **cell** is the smallest unit of living matter. When you looked at a layer of the onion leaf under the microviewer, you were able to see its cells. Look again at your notes. How did you describe a cell? How were the cells arranged? Could you see that their arrangement helps give structure to the leaf?

Cells differ in size and shape. Did you know that the yolk of an ostrich's egg is the largest cell? Some cells are so small you can only see them under a microscope.

***Ostrich—
largest of all living birds***

Ostrich egg

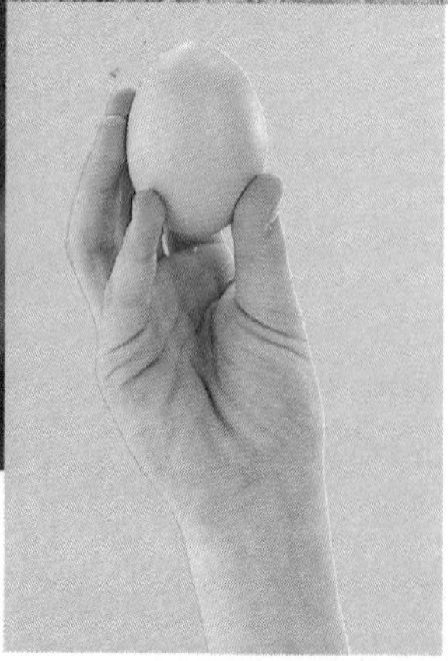

Chicken egg

We have been talking about the way systems work together to carry out the life processes of the organism. We've also seen that the organs work together in the system and the tissues work together to make up the organ. All these parts work together to help the organism function. Because the cell is the smallest unit of matter, the processes of life must be carried out by the cells.

One-Celled Organisms

In the activities you've done and in the examples you have read about, you were looking at and thinking about organisms that have many cells that work together. Are there any organisms that don't have many cells? How would they be able to live if they didn't have cells that were grouped into tissues, organs, and organ systems? Before studying about one-celled organisms, do the next activity to see some one-celled organisms in pond water.

Euglenas are one-celled organisms that live mainly in fresh water. They have properties of both plants and animals. They have structures for food production, but they also have structures for movement and taking in food.

An amoeba is another kind of one-celled organism that lives in water. It surrounds its food and breaks it down for use in its life processes.

Pond Organisms

Because one-celled organisms are so small, it's often useful to view them under a microscope. In this activity you'll look at a drop of pond water to see if you can find any one-celled organisms.

What You Need

microscope, pond water, slide, coverslip, dropper, *Activity Log* page 10

Place one drop of pond water on a slide and cover it with a coverslip. Observe the slide under the lowest power on your microscope. Look closely for organisms that are swimming in the water. How many cells make up the organism? Draw a sketch of one organism. Does it have any parts to aid its movement?

How Do They Move?

Amoeba move by *pseudopods*. After a pseudopod pushes out in the direction of its movement, the rest of the amoeba flows in the same direction. What do *pseudo* and *pod* mean? Use a dictionary to find the origin of the word. Then draw a picture in your ***Activity Log*** on page 9, of this movement of amoebas.

Focus on Technology

Spare Parts for Humans

You've seen how parts of a system are very important to the organism. If parts don't function properly, then carrying out the necessary functions of life is more difficult. Humans are able to use artificial parts to help them perform everyday tasks.

Minds On! Try not to use your thumbs for a minute. Write your name and tie your shoe. Did you find those two tasks hard? ●

Prostheses (pros thē′ sēz) are artificial body parts. They replace body parts that are missing at birth or that have been removed, which is called amputation (am′ pyə tā′ shən). Amputation may be necessary because of birth defects, diseases, or accidents. Prostheses have been used to replace amputated limbs since 300 B.C. Scientists have made them as close to real body parts as possible in appearance and in function.

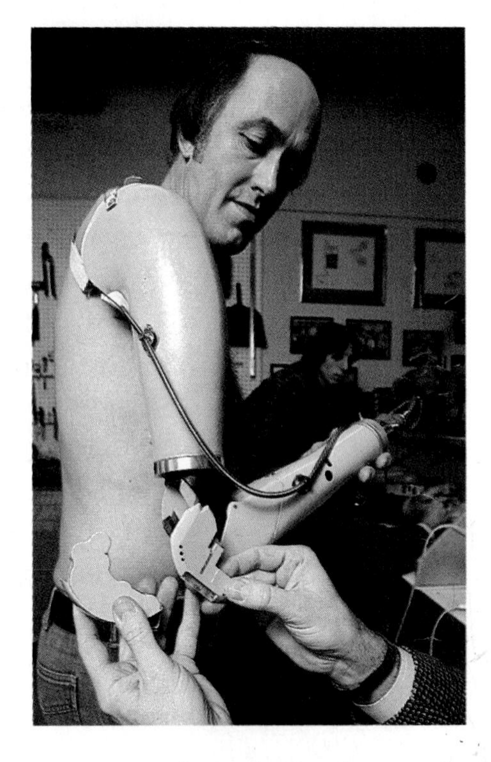

An example is the Utah artificial arm, developed by Dr. Stephen Jacobsen in 1981. It's for people with arms amputated above the elbow.

The Utah arm is battery operated. An electronic system connects the prosthesis to the upper arm.

The Utah arm can sense the electrical activity of the upper arm's muscles. These muscle signals are then used to move the artificial limb.

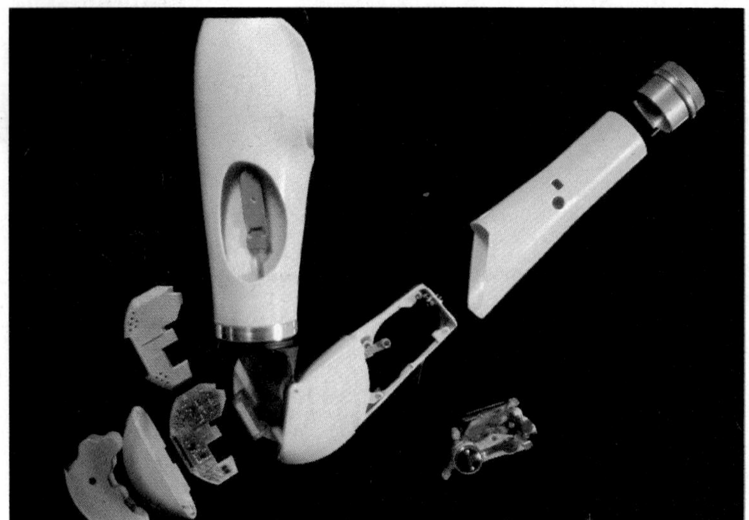

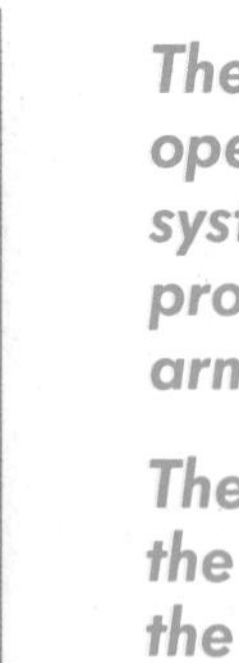

This system is so complex it enables a person to tie a shoe!

Prostheses can't be expected to work exactly like the parts they are replacing, but they can help those who need them lead more active lives.

Sum It Up

All organisms, whether they are one-celled or many-celled, perform certain similar functions to stay alive. In many-celled organisms, individual cells are at work to help with the processes of the organism. Many activities occur in the body of an animal or plant at any one time. The organism's cells work together in tissues, organs, and organ systems to carry out the organism's life processes. All the systems work together to keep the organism alive. If one system doesn't do its job, the entire organism is affected. The interaction of all the systems in an organism is necessary for life. One-celled organisms don't have tissues, organs, and organ systems, but they also carry out life processes.

Minds On! How are you different from plants, one-celled organisms, and many-celled organisms? Select one organism and in your ***Activity Log*** on page 11, draw a picture of its systems, organs, tissues, and cells based on what you know. ●

Critical Thinking

1. Humans are able to replace body parts with prostheses. How do plants replace parts that are missing?

2. How are an organism's systems like a transport system (buses, cars, trucks, etc.) in a city?

3. What plant and animal organs and tissues do you eat?

Macaw

Protozoa

HOW SMALL IS LIFE?

Look around. What's the smallest living thing you see? Can you think of other living things that are even smaller? What about a flea or a seed? Even these tiny things have smaller parts that you can't see without help. In this lesson, you'll look at these smaller parts that make up living things.

There is a whole miniature world that can't be seen with the unaided eye. Before magnifying lenses and microscopes were used, no one imagined such a world existed. However, when people began to look into this new world, they began to find answers to questions about organisms.

Minds On! Make a list in your ***Activity Log*** on page 12, of all the small living things you can think of. Choose one and draw a detailed, enlarged picture of it showing what you think it's made of. ●

When scientists looked into the world under a microscope, they discovered that the smallest unit of living matter was a cell. Then they looked even closer to learn what a cell was made of and how it worked.

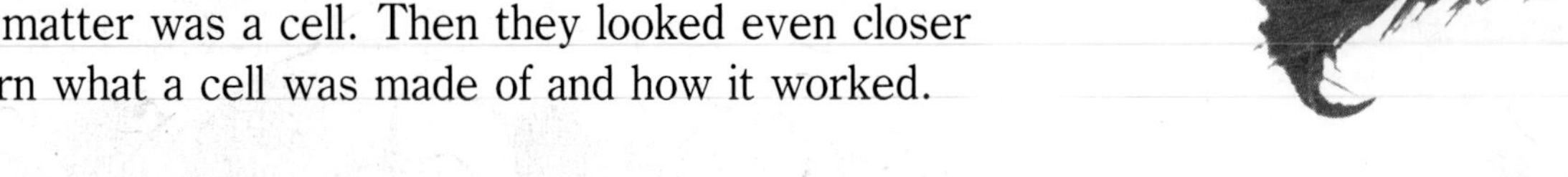

Asian ant on another Asian ant from the same nest

TRY THIS Activity!

How Small Is Small?

You can look even closer at a cell by looking through a microviewer.

What You Need

microviewer, microslide, *Activity Log* page 12

Compare the cell to the tiny living thing that you drew. How small is this cell? How can things move in and out of cells?

EXPLORE

Activity!

In and Out of Cells

What do you think a cell might need to stay alive? How does a cell get what it needs and get rid of what it doesn't need? Most kinds of cells are very small, too small to see their parts without a microscope. In this activity you'll examine a model of a cell to see some of the processes that enable it to live.

What You Need

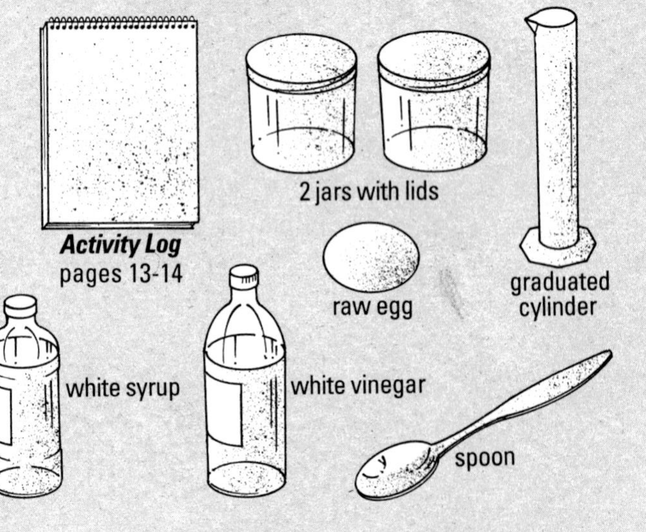

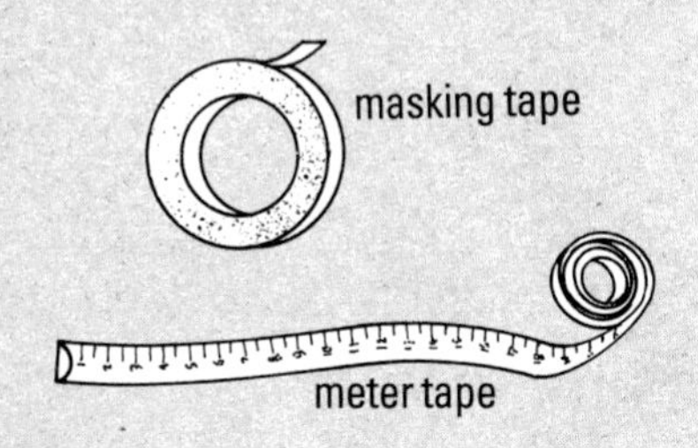

What To Do

1 Label the jars A and B. Measure 200 mL of vinegar into glass jar A.

2 Place the measuring tape around the center of the egg to find the circumference of the egg. Record the measurement in your ***Activity Log.***

3 Carefully place the egg into the vinegar. Place the top on the jar and leave it for 2 days.

4 After 2 days, remove the egg from the vinegar with a spoon and carefully rinse it with water Use the measuring tape to find the circumference of the egg. Measure the vinegar remaining in the jar with the graduated cylinder. Record your measurements in your ***Activity Log***.

5 Measure 200 mL of syrup into jar B. Place the egg into the jar of syrup. Place the lid on the jar and leave it for 1 day.

6 Carefully remove the egg with the spoon from the syrup. Rinse it with water. Use the measuring tape to find the circumference of the egg. Measure the syrup remaining in the jar. Record your measurements in your ***Activity Log***.

What Happened?

1. What happened to the shell of the egg?
2. Vinegar is made of acid and water. Which part dissolved the shell? What was left to move into the egg?
3. What happened to the size of the egg after remaining in vinegar?
4. What happened to the size of the egg after remaining in syrup?

What Now?

1. What part of the egg controlled what could move in and out of the egg?
2. Predict what would happen if the egg removed from the syrup were placed into water. Now try it. Explain what happens.

The Cell

In the Explore Activity, you saw a chemical reaction between the eggshell and the acid in the vinegar. The acid caused the shell to dissolve. The water in the vinegar moved into the egg, and the size of the egg increased. The water moved into the egg because there was a greater percentage of water molecules outside the membrane of the egg than there was inside. The water moved out of the egg when it was placed in the syrup because there was a greater percentage of water molecules inside the membrane of the egg than there was outside.

All organisms are made of cells. A cell is the smallest unit of an organism that is capable of life. Cells are made of atoms, the smallest units of all matter.

Cells of organisms differ, and there are many different shapes of cells. Bacteria and some fungi have only one cell. Some organisms, like humans, have trillions of cells.

Chromosomes *(krō´ mə sōmz´) are cell parts inside the nucleus that carry the information that determines the characteristics an organism has. Chromosomes determine your eye and hair color and make you different from other people.*

The ***nucleus*** *(nü´ klē əs) is a dense, dark structure near the center of the cell. It controls the activities of the cell. It functions much like the brain that controls the activities that go on in the cell.*

A **cell membrane** (mem´ brān) surrounds the cell just as skin covers your body. The cell membrane holds the other parts of the cell in place and protects the cell. It allows water and other molecules to pass into and out of the cell. It also prevents the passage of certain molecules in and out of the cell.

Inside the cell membrane is a jelly-like substance called **cytoplasm** (sī´ tə plaz´ əm). The cytoplasm is made of water and other chemicals. The nucleus and other cell parts are located in the cytoplasm. Chemical reactions take place in the cytoplasm.

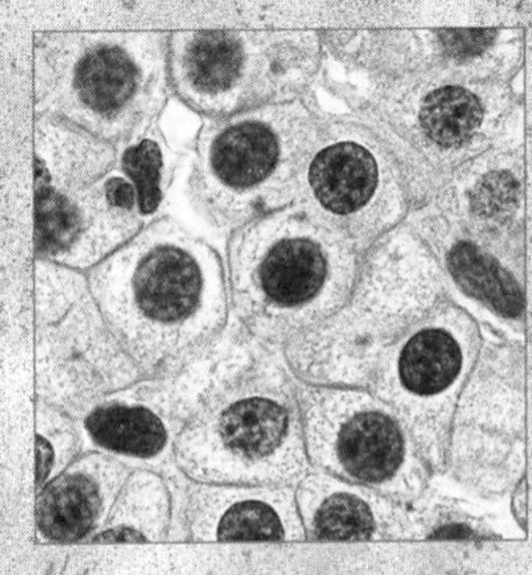

Liver cells

There are other structures that make food, release energy for the cell to use, and transport materials throughout the cell. All the cell parts work together to maintain the life of the cell.

In order to live, all organisms, even one-celled ones, do certain things. All living things reproduce, grow, develop, need food, use energy, and respond to stimuli. They need to get energy and use it to grow and reproduce. They respond to stimuli. For example, if your stomach makes sounds and feels empty, you eat food. Your hunger is the stimuli that causes you to eat and helps you survive. These things enable organisms to survive.

How Are Plant and Animal Cells Different?

Plant cells are different from animal cells. Plants have needs that animals don't have and cell parts that meet these needs.

PLANT CELL

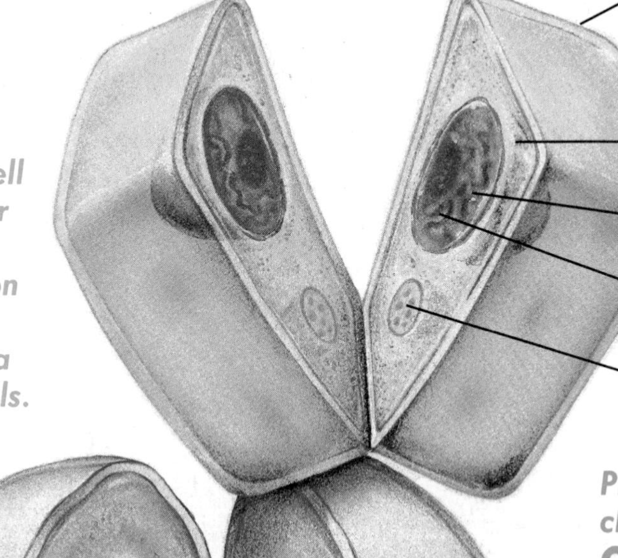

*A plant cell has a **cell wall** that surrounds the cell membrane. The cell wall is a stiff outer covering that provides protection for the cell. Fungi and some bacteria also have cell walls.*

ANIMAL CELL

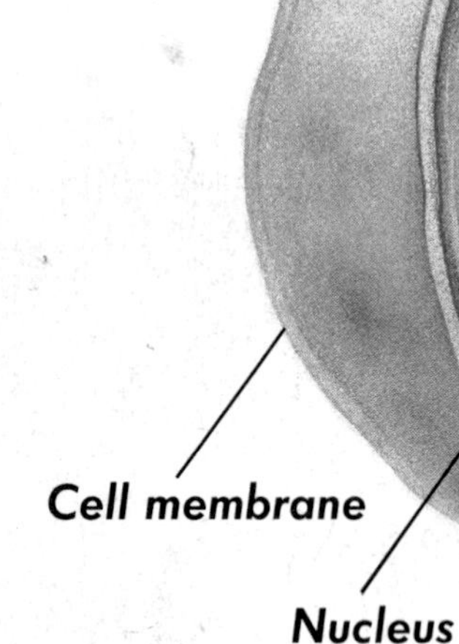

*Plant cells have chloroplasts. **Chloroplasts** (klôr´ ə plasts´) are cell parts that contain chlorophyll. Chlorophyll is a green-colored material that gives plants their color. It is used in the process of photosynthesis (fō´ tə sin´ thə sis) to make food for the plant.*

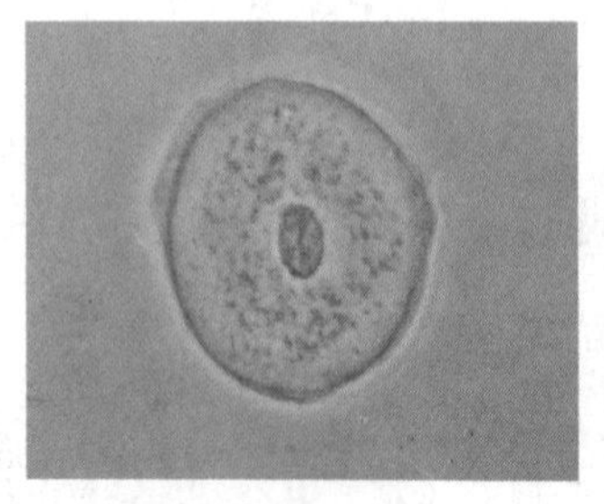

Typical animal cell

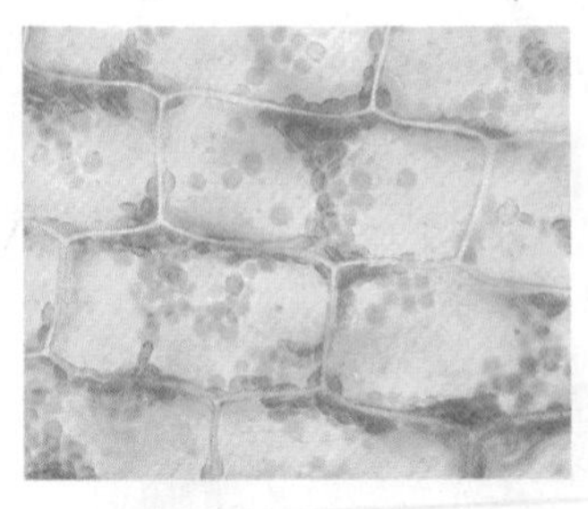

Typical plant cell—nuclei not seen from this view

Language Arts Link

Little Rooms

In 1665 Robert Hooke, an English scientist, observed a slice of cork through a microscope. He saw that it was made up of small units that fit together and looked like a honeycomb. He named these "little rooms" cells. What he actually saw were the cell walls of the cork plant.

Look at the cork cells in the photograph below with a hand lens. Describe what you see. Draw a picture in your ***Activity Log*** on page 15 of the cork cells. Label the parts you can identify. Imagine that you are Hooke. Use your notes to write a letter to another scientist describing your discovery. Consider the questions that Hooke might have had and that he might have asked of his fellow scientists.

TRY THIS Activity!

Animal and Plant Cells

In this activity you'll observe a plant cell and an animal cell and determine how they are different.

What You Need

microviewer, microslides of onion skin and human cheek cells, *Activity Log* page 15

Look at each of the slides under the microviewer. In your ***Activity Log***, sketch what you see in each cell. Label the cell parts. How are the onion skin cells and cheek cells alike? How are they different?

Corks

Cork cells

Cork oak tree

Microscopes

One of the most important tools used by scientists is the microscope. You can make a simple one yourself. Put a penny under a glass of water and look at it. What happens to the penny? Engravers used water-filled globes as magnifying glasses as long as 3,000 years ago. Glass lenses weren't used until the late 1200s. A Dutch maker of reading glasses, Zacharias Zanssen (zän′ sən), put two magnifying glasses together in a tube. He is credited with discovering the principle of the compound microscope about 1590. Then in the mid 1600s, Anton van Leeuwenhoek, (lā′ vən hük) another Dutch scientist, made microscopes that could magnify up to 270 times. He was one of the first people to record what he saw under the microscope.

A light microscope lets light pass through the object being looked at and then through two or more lenses. An image can be magnified up to 2,000 times.

An electron microscope uses a magnetic field to focus beams of electrons. An electron microscope can magnify up to 1,000,000 times.

The two kinds of electron microscopes are the transmission electron and scanning electron. The transmission electron microscope is used to study the inside parts of the cell. It can magnify up to 300 thousand times but it has one disadvantage. The object to be looked at must be sliced very thin. Only dead cells can be observed.

The scanning electron microscope is used to observe the surfaces of whole cells. Its magnifying power isn't nearly as great, but it's possible to see a more realistic view with it.

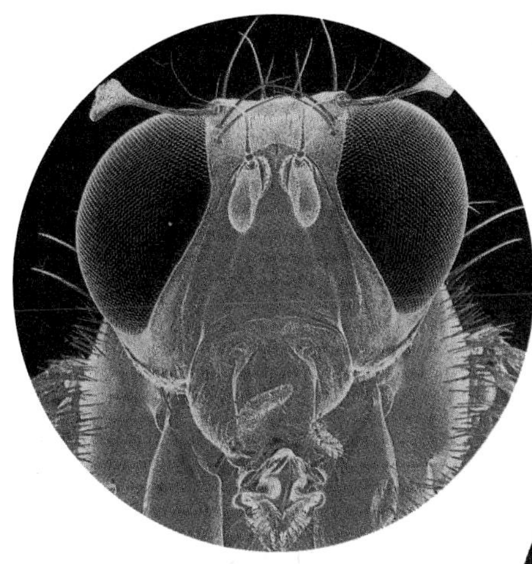

Fruit fly

House dust mite

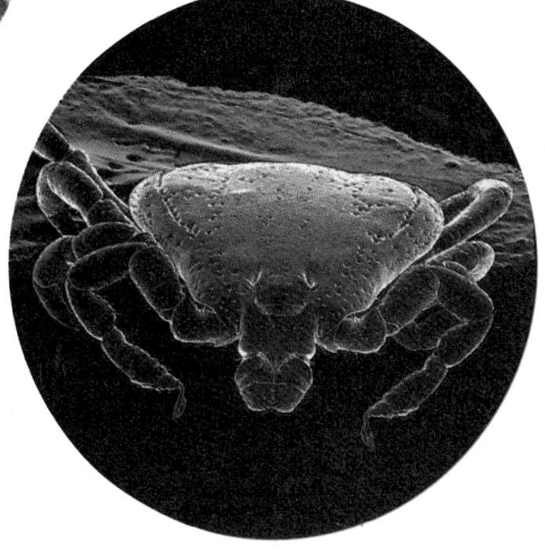

American dog tick

Sum It Up

Scientists hypothesize about the number of organisms in the world. Under a microscope you can see that organisms are made of cells. The smallest organisms have only one cell. Some organisms are made of trillions of cells. All cells have parts that work together to carry out life functions. The nucleus controls the activities of the cell. The chromosomes carry characteristics to the offspring. The cell membrane allows certain materials to pass through. All the life activities of the cell take place in the cytoplasm. All these parts and others help the cell perform its functions to live.

Ragweed pollen

Critical Thinking

1. Why is it important for the cell membrane to allow some substances to pass into and out of the cell and prevent others from passing?
2. Why are plant cells different than animal cells?
3. How are microscopes helpful to the field of medicine?

Magnifications vary

Fungus

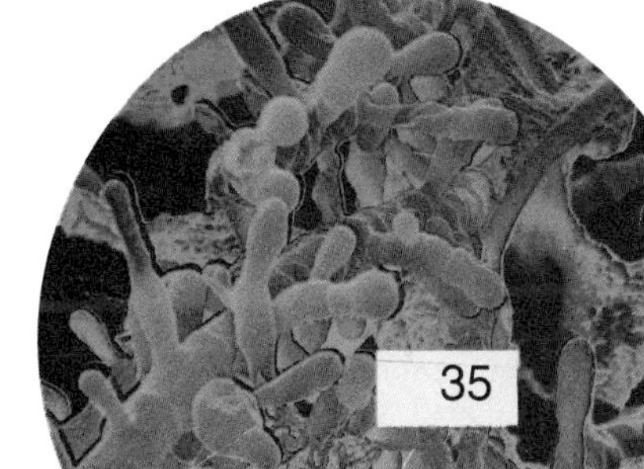

HOW DOES LIFE REPLACE ITSELF?

Have you ever seen baby birds hatching from their eggs? Have you ever planted seeds and watched them sprout? New life is exciting. Each day there is growth and change. Living things must reproduce for life to continue. But how does the process of reproduction take place?

Cheetah and cubs

You know that when you cut your hair or your fingernails, they grow back. You may have planted seeds and watched them grow. Do you think all living things grow and reproduce in the same way?

Minds On! Imagine being in a park, noting all the plants and animals. What would happen if there were no new organisms to replace the old ones? How would it affect the park? ●

Math Link

My, How You've Grown!

Find out how much you weighed at birth. Did you grow steadily or in spurts? In your ***Activity Log*** on page 16, subtract that weight from what you weigh now. Divide that number by your age to find the average weight you've gained each year.

Though different organisms may carry out their activities in different ways, all the processes that go on allow them to survive. How do you think plants grow and develop?

EXPLORE Activity!

How Are New Plants Made Without Seeds?

You know that new plants can be produced from seeds. In this activity, you're going to look at other ways plants can be produced.

What You Need

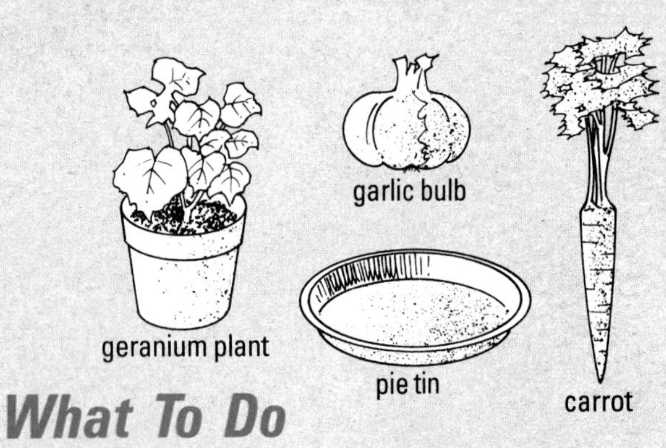

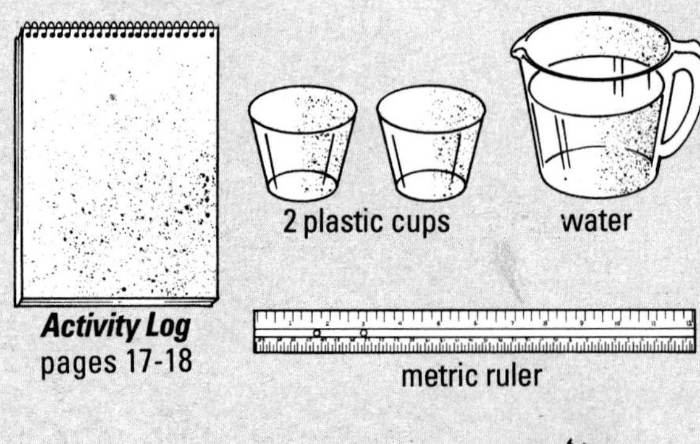

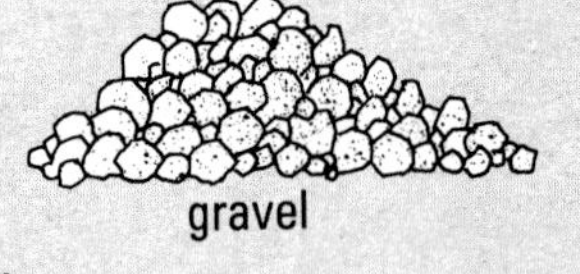

What To Do

1 Cover the bottom of the cup with gravel. Finish filling it with moist soil.

2 Break the garlic bulb apart. Plant 2 cloves about 3 cm deep in the soil. Keep the soil moist but not wet.

3 Use a metric ruler to measure about 2 cm from the top of the carrot. Have your teacher cut the bottom off the carrot for you. Place the cut side down in the pie tin filled with water. Keep water in the dish.

4 Cut a small branch close to the stem from a geranium plant. This is called a cutting. Place the cutting in a cup of water.

5 Predict what will happen to each plant.

6 Check all 3 plant parts every day. Make a drawing of what you see every 3 days in your ***Activity Log***.

What Happened?

1. Describe what happened to each of the plant parts.
2. How many parents were needed for reproduction?

What Now?

1. What do you think causes something to grow?
2. The carrot plant will eventually produce seeds. How could this be helpful to a person who had one carrot?

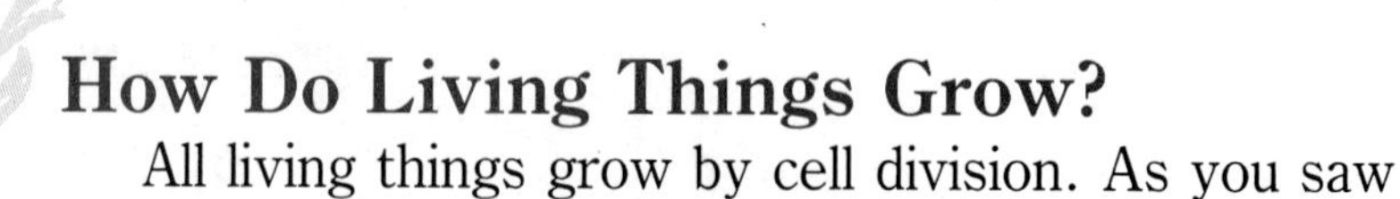

How Do Living Things Grow?

All living things grow by cell division. As you saw in the Explore Activity, the plants grew new parts. The new parts grew because plant cells were dividing. The plants produced new cells to increase their size.

These plants normally grow from seeds. In the Explore Activity, they grew from other plant parts. They produced a new organism like themselves if they kept growing.

Organisms produce new organisms similar to the parents through **reproduction** (rē´ prə duk′ shən). Reproduction may require one or two parents. Reproduction with one or two parents takes place through cell division. You watched new garlic, carrot, and geranium plants, just like their parent plants, grow from a bulb, a root, and a stem. These plants produced new cells to grow. In humans new cells are formed through cell division. One cell grows and divides into two. The cell can grow only so big, and then it divides.

Cells are dividing to produce leaves on the top of the carrot.

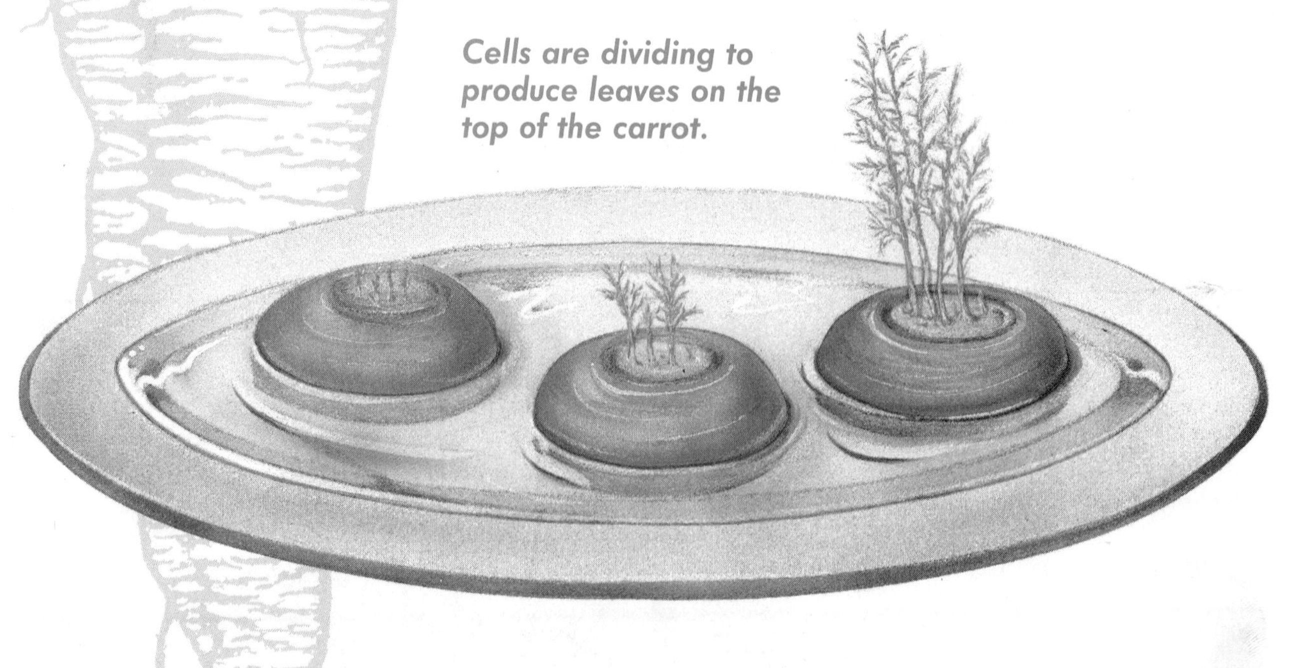

Eggs

Tadpole

Immature frog

Adult frog

In many-celled organisms, the cells separate but don't form two new organisms. One cell divides into two. Two cells divide into four, and so on. The organism grows in size as it adds more cells.

In one-celled organisms like the paramecium, one cell separates to form two new organisms.

The organism begins and ends its life as a single cell.

Mitosis and Asexual Reproduction

Cell division that produces cells just like the parent cell is called **mitosis** (mī tō′ sis). Some one-celled organisms, such as the paramecium, reproduce by mitosis. Do the activity to learn more about cell size.

MITOSIS IN A PLANT CELL

1. At the beginning of mitosis, chromosomes are in the nucleus.

2. The chromosomes in the nucleus have made a copy of themselves.

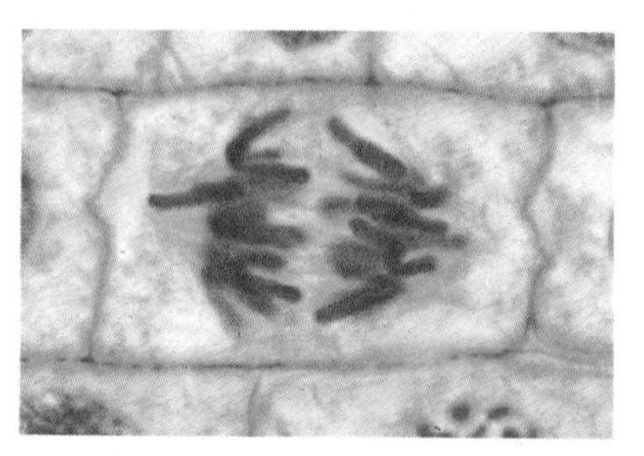

3. Then the chromosomes separate.

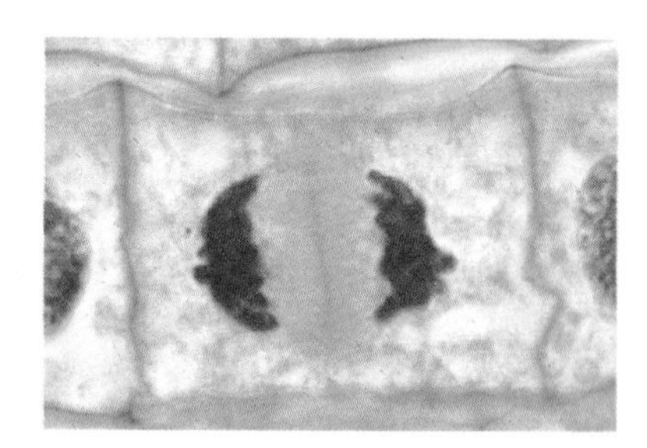

4. The cell divides. One set of chromosomes goes to each of the new cells.

TRY THIS **Activity!**

How Many Cells?

How many cells are under a one-centimeter square?

What You Need

metric ruler, sheet of white paper, scissors, *Activity Log* page 19

In your ***Activity Log***, predict how many cells are under a 1-cm square. With a metric ruler, draw a square 1 cm on each side of a sheet of paper. Draw 150 pencil dots on the square. Use your scissors to cut out the square. Place the square on your hand. Under the square are more than 150 thousand skin cells, most of which will be gone tomorrow. What will happen to the old skin cells from your hand?

Most of the cells in many-celled organisms divide by mitosis. For example, skin cells reproduce by making exact copies of themselves. As old cells wear out and die, they're replaced by new cells. What factors contribute to the healthy growth of organisms?

In other parts of your body, other types of cells are also reproducing.

Every second, over two million red blood cells die and are replaced.

Asexual (ā sek′ shü əl) **reproduction** requires only one parent. Some organisms reproduce through mitosis by splitting in half or by outgrowths from a parent cell. Amoebas, paramecia, yeasts, and other one-celled organisms can reproduce through asexual reproduction. Two new organisms form from the division of one parent cell.

TRY THIS Activity!

Yeast Life Processes

What do yeast cells need to reproduce?

What You Need

dry yeast, spoon, sugar, plastic cup, warm water, *Activity Log* page 19

Add 1 tsp. of dry yeast to the cup that contains 1 Tbsp. of sugar and 9 Tbsp. of water. Observe the cup. What happens in the cup?

The yeast cells use the sugar for food. The food allows them to grow and reproduce. Yeast cells reproduce by outgrowths from the parent cell.

Meiosis and Sexual Reproduction

In organisms that have two parents, another type of cell division takes place. **Meiosis** (mī ō′ sis) produces cells that have only half the number of chromosomes of the parent-body cell. The diagrams on these pages show the results of meiosis and sexual reproduction in animals.

Fruit flies

*In females, sex cells formed by meiosis are called **egg** cells.*

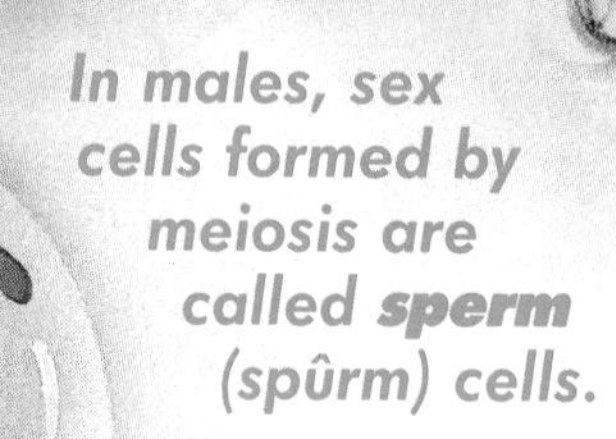

*In males, sex cells formed by meiosis are called **sperm** (spûrm) cells.*

Sexual (sek′ shü əl) **reproduction** is reproduction in which sex cells join to form a cell that has the same number of chromosomes as the parents.

*The cell formed by the joining of the sperm cell and the egg cell is called a **zygote** (zī´ gōt).*

Because an equal number of chromosomes have come from each of the parents, characteristics from both parents are found in the zygote.

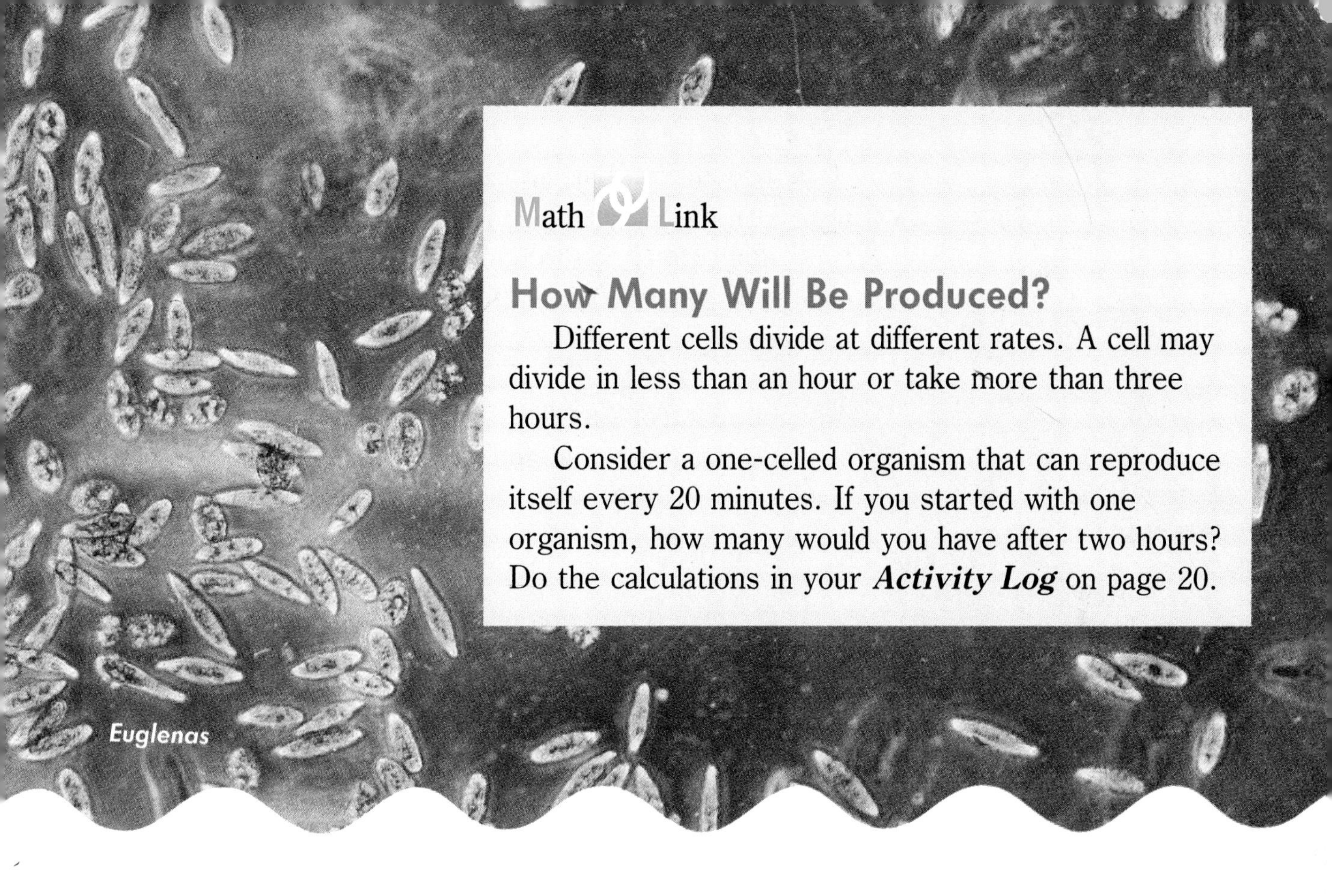

Math Link

How Many Will Be Produced?

Different cells divide at different rates. A cell may divide in less than an hour or take more than three hours.

Consider a one-celled organism that can reproduce itself every 20 minutes. If you started with one organism, how many would you have after two hours? Do the calculations in your ***Activity Log*** on page 20.

Euglenas

As the cells divide, some cells change into cells that may become nerve cells, muscle cells, or skin cells. The new organism has parts and characteristics of both parents.

*The zygote divides by mitosis to form an **embryo** (em´ brē ō´). The cells in the embryo grow and divide by mitosis.*

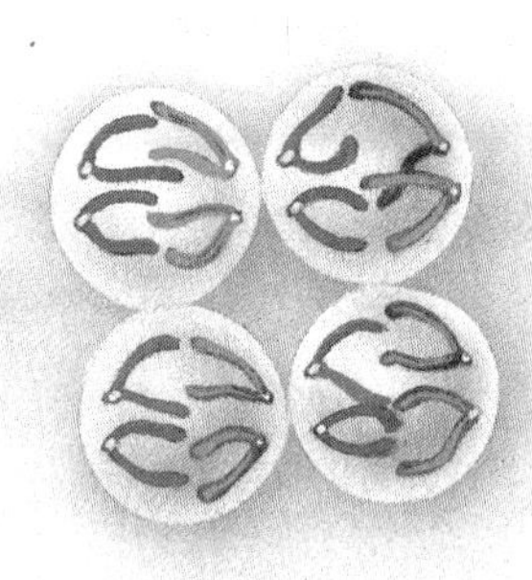

Horticulturist

A horticulturist grows, markets, and cares for plants. Horticulturists change the reproductive cycle of plants to fit the needs of the buyers. They can control the blooming of flowers using various techniques. Many of the plants they grow are started from plant cuttings, like the geranium in the Explore Activity on pages 38–39.

As a young child, Emily Ketter loved to work in the garden with her father. Ms. Ketter thinks that these early experiences may be what first interested her in pursuing a career in horticulture.

A horticulturist may teach about plants at a university or study ways to eliminate plant disease. A horticulturist may work in a greenhouse, on a farm, in a garden center, in a park, or in a research lab. The horticulturist usually has either a two-year or a four-year college degree. It's possible to obtain an entry-level job with a high school background along with practical experience.

Ms. Ketter says, "In order to have a career in horticulture, students need to train their minds in math and science. To be successful, though, they also must truly love to work with plants. I know that I wouldn't want to do anything else."

Through technology in biology, people have learned to control the growth and reproduction of many organisms. Do you think this is right? What do you think of its use for medicine?

Technological Advances With Living Things

Farmers making plans for next year's crops need to know how many seeds to buy. Those raising livestock have to know how many animals their cows, sheep, and pigs will produce. An understanding of the processes of growth and reproduction is important to people in many careers.

Dairy farm, Pennsylvania

Animals Used for Producing Medicine

Scientists have changed the chromosomes in sheep and goats so they produce drugs we can use in medicine. The medicine can be used to help sick people. Some people don't think scientists should change animals this way. The controversy is whether the benefits of producing medicines in this manner give humans the right to alter the chromosomes of animals. How do you feel about this process? Write a paragraph in your ***Activity Log*** on page 21 telling how you feel and giving reasons for your position.

Sheep

Focus on Technology

"Living Stamps"

Within the last ten years, there have been exciting technological breakthroughs that are being applied in the treatment of burn victims. These breakthroughs have provided a way to heal severe burns that cover most of the patient's body in as quickly as 21 days.

Technicians surgically remove skin tissue the size of a postage stamp from a part of the body that has not been burned.

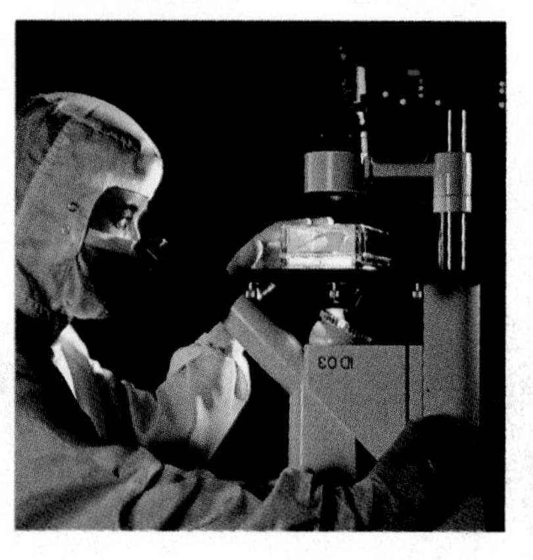

Seen under a microscope, as many as six million skin cells have multiplied. These cells are separated and placed into six glass dishes, each glass dish holding about one million independent cells.

Putting the cells into the dishes is the first stage of the process. The cells multiply in dishes for ten days by undergoing mitosis. By the end of the first stage, the original number of cells has multiplied by 70!

During the second stage, the sheets of cells that have formed are separated into individual cells again and put into new dishes. By the 21st day, the cells have multiplied in number by 10,000, have grown together, and are ready to be applied to the patient's wounds. As early as six days after application to the wounds, new skin tissue forms from the cells. Tests of this new skin have shown that it has many qualities of skin that has never been burned.

Sum It Up

Organisms grow and reproduce so that life can continue. Organisms reproduce by sexual reproduction or asexual reproduction. All organisms' cells reproduce by mitosis. Meiosis is cell division that occurs in organisms that reproduce with two parents. Meiosis produces sex cells that join to form a zygote, then an embryo. The embryo develops and matures with characteristics of both parents.

Minds On! Looking at old pictures of yourself, you see that changes have occurred as you've grown. Even when fully grown, some of your cells will continue to divide and reproduce to replace worn-out cells. As new cells replace old ones, the person you see in the mirror is a "new" you. Imagine what changes you have to look forward to. ●

Family looking at pictures

Critical Thinking

1. What would happen if mitosis did not occur in humans? Why? What would happen if meiosis did not occur? Why?

2. Explain why reproduction is important for the survival of the species.

3. Damaged skin can be replaced by generating new skin from cell tissue. What other applications might be developed from this method?

LIVING KINGDOMS

There may be millions of living things in the world that we know nothing about. If you discovered a living thing, what would you name it? How could you tell people about it? How would you keep track of all living things? In this lesson, you'll learn what questions scientists ask to group or classify organisms.

Mountain goats

Soil bacteria

Protozoa

Mushroom

Yellow lotus

Peregrine falcon

Cactus

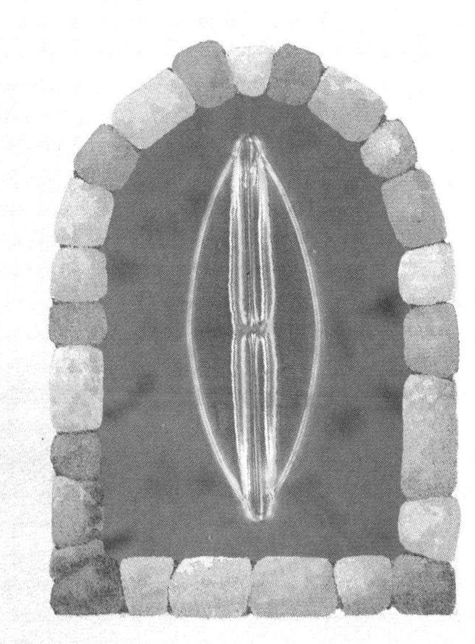

Diatom

Minds On! Look at the organisms on this page. How would you tell them apart? How would you describe them? In your ***Activity Log*** on page 22, write down the characteristics you might use to classify them. How many groups did you make?

We tend to think of living things as either plants or animals. Do all organisms fit into these two categories? What about insects? Are they animals? What about the mushrooms you use in your salad or the algae floating in your fish tank?

In the next activity, you'll look at cells belonging to each of the five groups of living things. You will see how cells are used to classify organisms. What characteristics are used to group them together?

EXPLORE

Activity!

How Are Cells Used in Classification?

You know that every cell is made of cytoplasm and is surrounded by a cell membrane. Some cells have a nucleus. Some have other cell parts. Some have a cell wall. In this activity you'll examine cells from different organisms and observe how they are alike and different.

What You Need

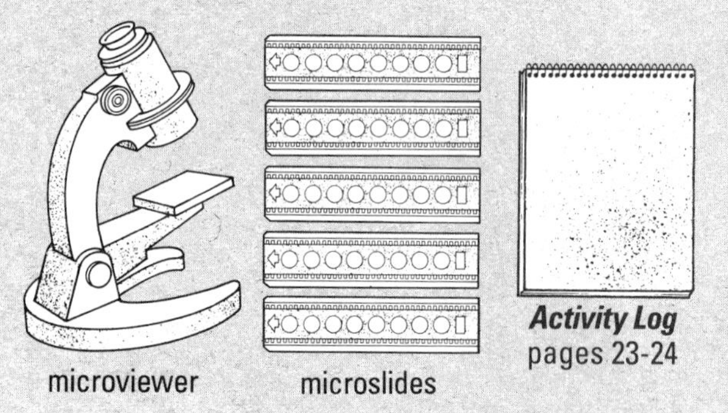

What To Do

1 Look at each of the cells carefully through the microviewer.

2 After looking at each cell, draw a picture of the cell and record the cell parts that you see in the data table in your ***Activity Log.***

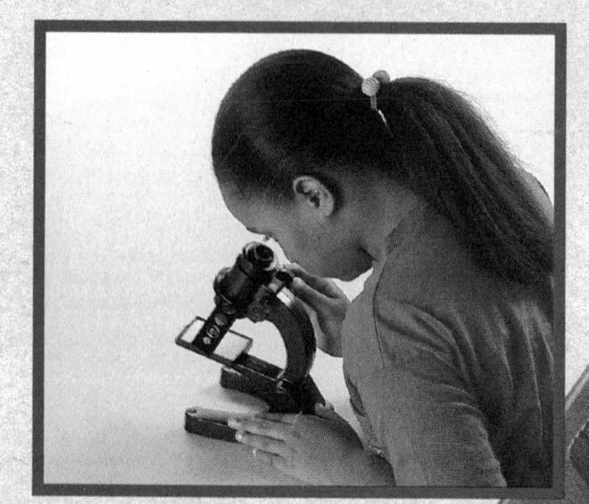

3 When you have finished observing all of the cells and recording your observations, use your data to answer the following questions.

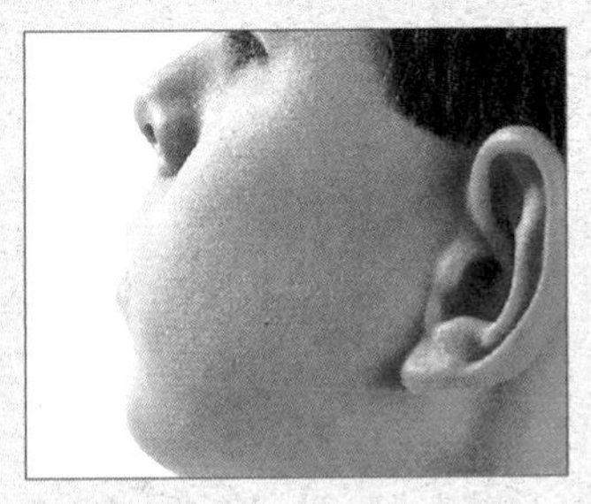
Human cheek

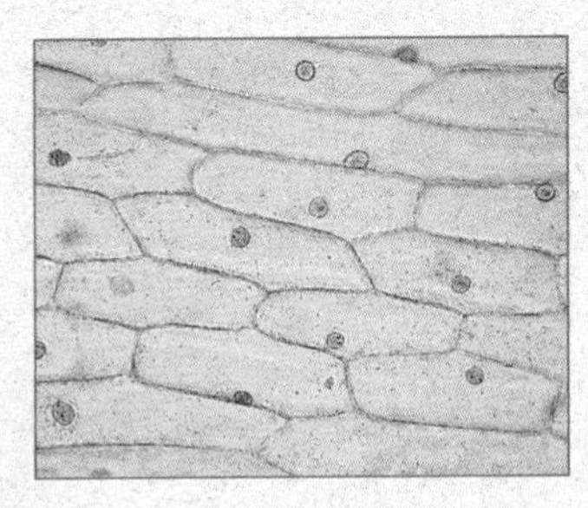
Onion skin

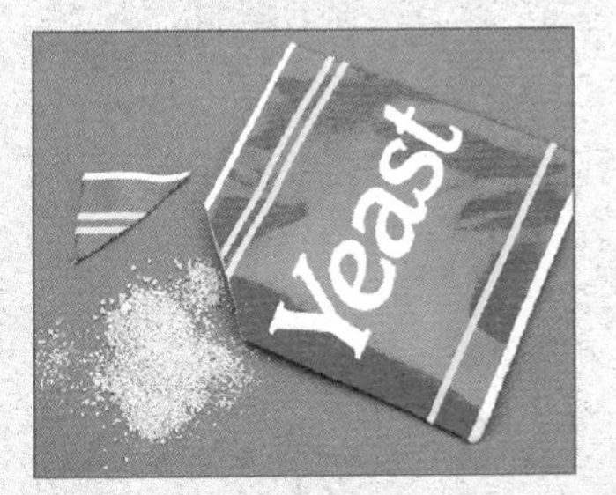

Yeast

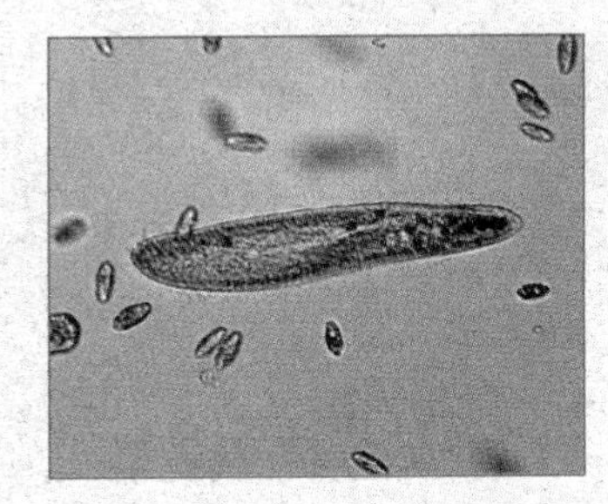
Paramecium

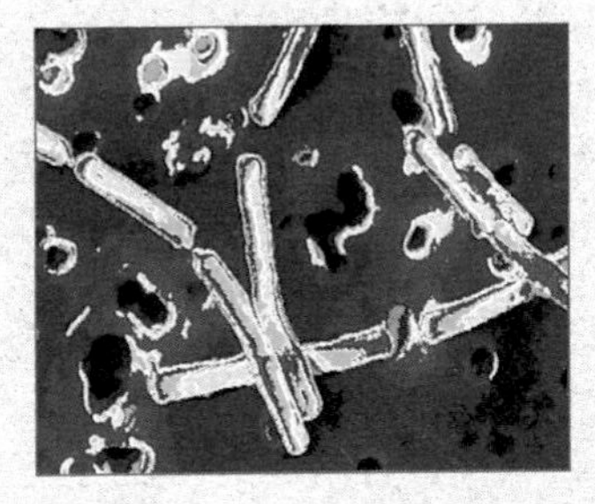
Bacteria

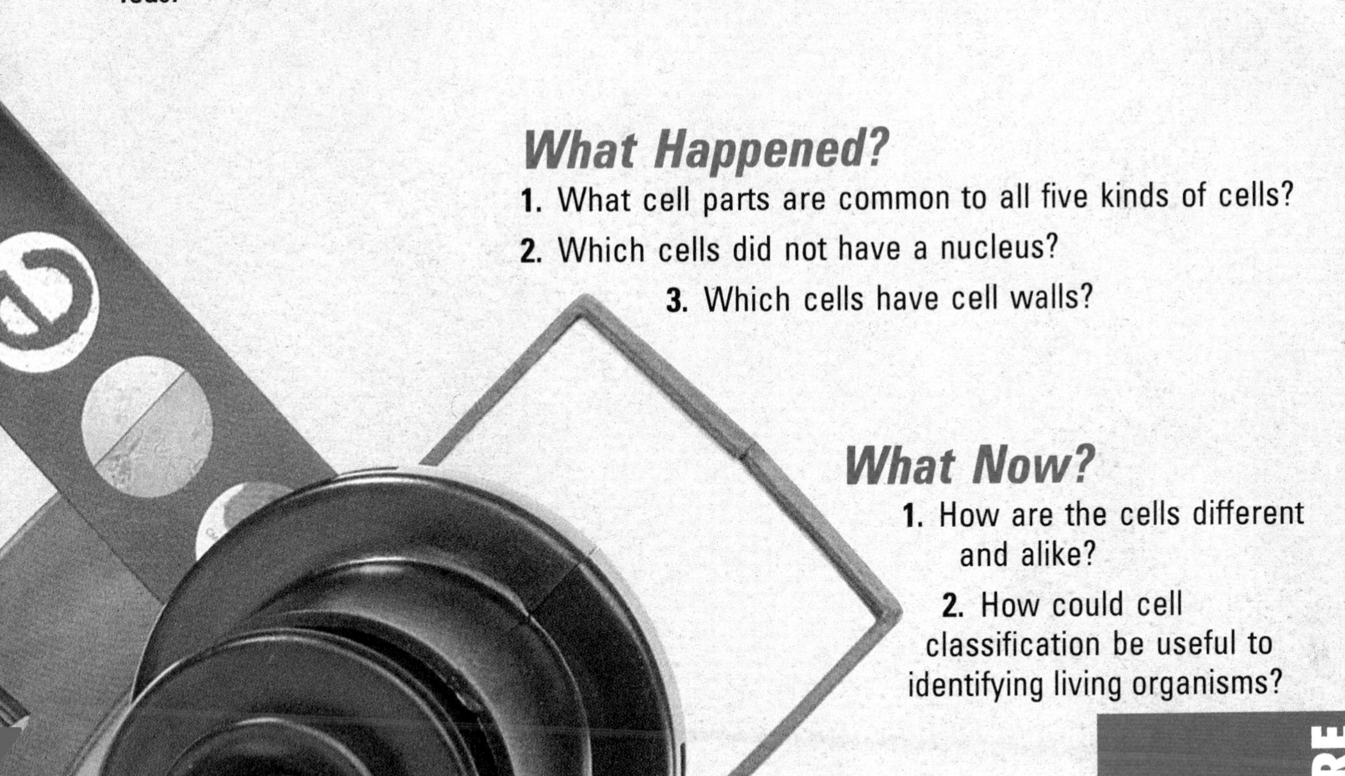

What Happened?

1. What cell parts are common to all five kinds of cells?
2. Which cells did not have a nucleus?
3. Which cells have cell walls?

What Now?

1. How are the cells different and alike?
2. How could cell classification be useful to identifying living organisms?

The Five Kingdoms of Classification

Early scientists thought every living thing was either a plant or an animal. Over time more scientists found organisms that were closely related to plants and animals. This led to the method of grouping used by many scientists today. There are five groups.

A **kingdom** is the largest group of living things. The organisms of a kingdom are similar to one another but different from organisms in the other kingdoms. You looked at a cell from each of the kingdoms in the Explore Activity.

The five kingdoms are plant, animal, fungus (fung′ gəs), moneran (mə nîr′ ən), and protist (prō′ tist).

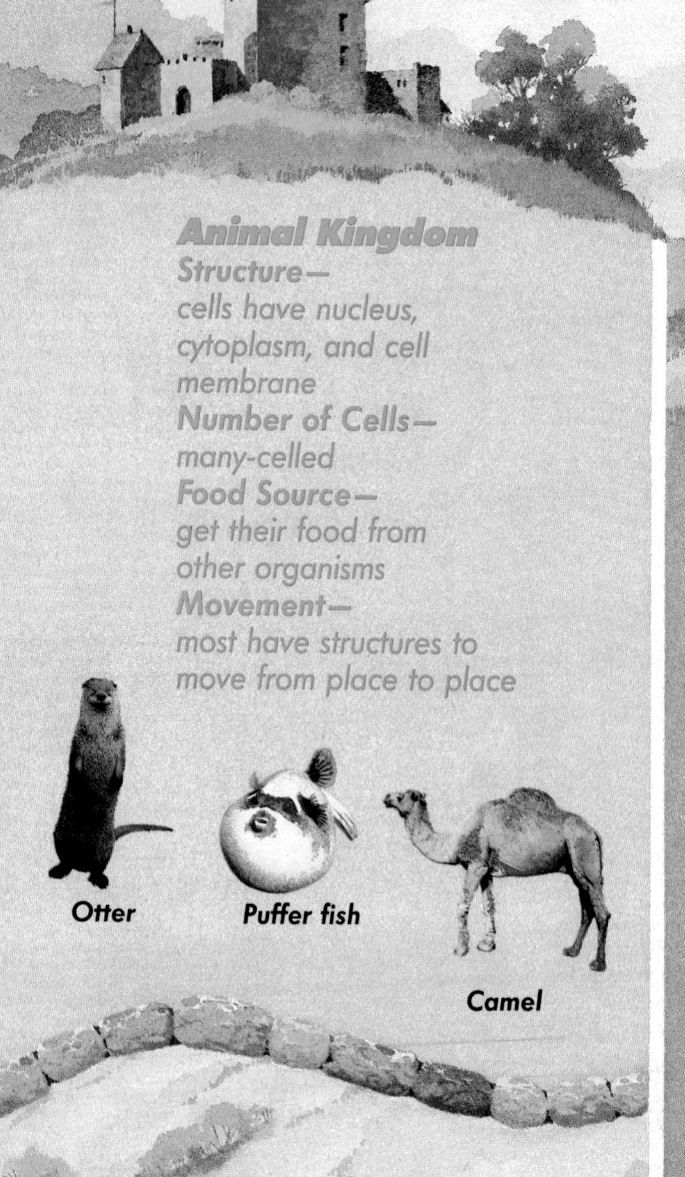

Animal Kingdom

Structure—
cells have nucleus, cytoplasm, and cell membrane

Number of Cells—
many-celled

Food Source—
get their food from other organisms

Movement—
most have structures to move from place to place

Otter

Puffer fish

Camel

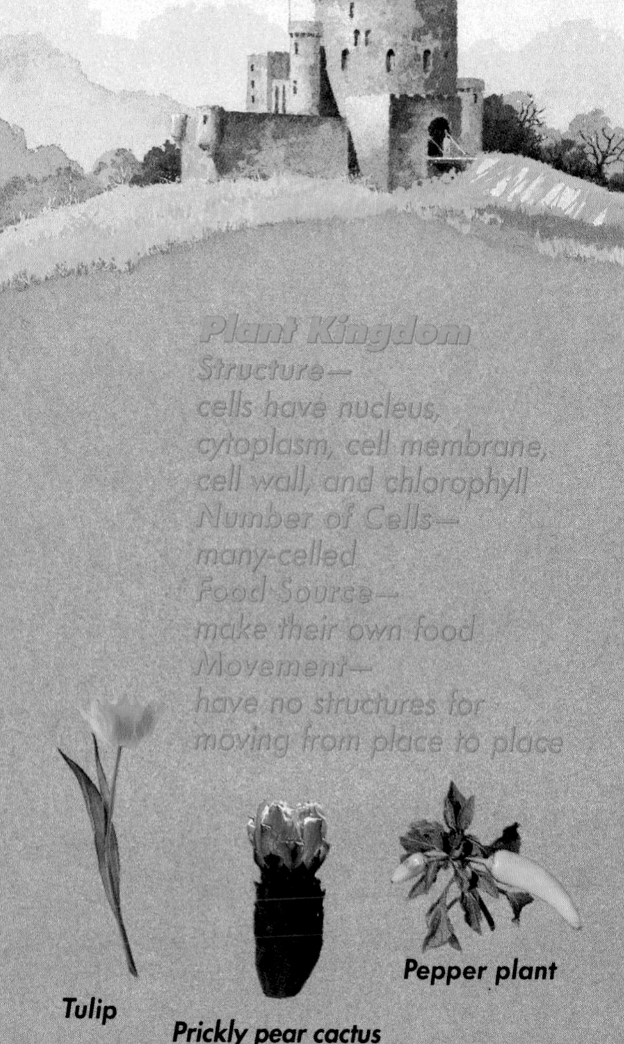

Plant Kingdom

Structure—
cells have nucleus, cytoplasm, cell membrane, cell wall, and chlorophyll

Number of Cells—
many-celled

Food Source—
make their own food

Movement—
have no structures for moving from place to place

Tulip

Prickly pear cactus

Pepper plant

Moneran Kingdom

Structure–cells have cytoplasm, cell membrane, and some have cell walls
Number of Cells–
one-celled
Food Source–some make their own food and some get food from other organisms
Movement–
some have structures for moving from place to place

Fungus Kingdom

Structure–
cells have nucleus, cytoplasm, and cell wall
Number of Cells–
many-celled, some one-celled
Food Source–
absorb their food from other organisms
Movement–
have no structures for moving from place to place

Protist Kingdom

Structure–
cells have nucleus, cell membrane, and cytoplasm
Number of Cells–
one-celled
Food Source–
make their own food or get food from other organisms
Movement–
some have structures for moving from place to place

Brown bear

Classification

When you looked at the cells under the microviewer, you saw that they shared common characteristics. You also saw that they were different from one another in some ways.

Scientists classify organisms by the characteristics of their cell or body structures. **Classify** means to group. After close study scientists can group some organisms with like characteristics. Scientists study to see if the organisms are alike because of evolution from one ancestor. **Evolution** is the change that occurs to organisms through time. If the organisms are similar because of evolution, they can be grouped together.

For example, to group organisms into kingdoms, scientists ask five questions:

1. Is the organism one-celled or many-celled?

2. Do the cells of the organism have a nucleus?

3. What cell parts do the cells have?

4. How does the organism get its food?

5. Can the organism move from place to place?

Minds On! Look back to the Explore Activity on pages 52–53. Using the information from the previous pages and the five questions above, determine which kingdom each of the cell types you looked at belongs to. Write the characteristics that helped you decide in your ***Activity Log*** page 25.●

Kingdoms are divided into smaller groups called phyla. A **phylum** (fī´ ləm) is the largest group within a kingdom. A **class** is the largest group within a phylum. Classes are divided into **orders.** Orders are divided into **families**. Families are also divided into smaller groups. Each of these groups is called a **genus** (jē′ nes). A genus is divided into the smallest and most closely related group, **species** (spē′ shēz). The seven main groups in scientific classification help keep all Earth's organisms in order.

Music/Art Link

New Organisms

Does a starfish look like a fish? Does a sea lion look like a lion? What other animal or plant names can you think of that are confusing or misleading? Make up an organism and draw a picture of it in your ***Activity Log*** on page 26. Have four of your classmates name your organism.

Sea horse

White horse

Sunfish

Starfish

Land crab

Crab apple

Why Do Scientists Name Organisms?

Swedish botanist Carolus Linnaeus (lə nē′ əs) (1707–1778), developed a system for naming organisms still used today. He gave each organism a two-word Latin name. These names were a combination of the genus and species names. They are used by people all over the world. When scientists around the world need to talk about living things, it's important that the names they use are understood by everyone.

Linnaeus developed his system before people understood evolution. Scientists revise the system to reflect new information. The changes they make help other scientists to understand the evolutionary relationships among living things.

Language Arts Link

What Did He Call Them?

Carolus Linnaeus wrote detailed descriptions of the organisms and found these descriptions helped him identify and name organisms.

Look back at your drawing of your new organism. Make up a scientific name based on its characteristics. Have four classmates make up scientific names for your new organism. Record the names in your ***Activity Log*** on page 26. How do they compare? Decide which name best fits your organism.

Dr. Blackwell from the United States viewing *Rana pipiens'* blood

Dr. Santiago from Venezuela viewing sangre de *Rana pipiens*

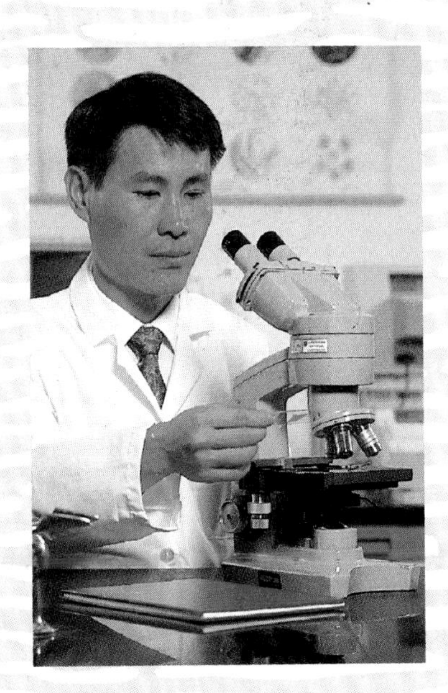

Dr. Xie from China viewing *Rana pipiens* 的血

Identifying Organisms

About a million and a half of all living things on Earth have been classified. As other organisms are discovered, other kingdoms may be needed. Classification systems have been proposed that have more than five kingdoms. Some people already use a classification system with seven kingdoms.

You can also use the five kingdom system to identify organisms. The system is based on evolved characteristics that organisms have in common. You can observe an organism's characteristics. When you know its characteristics, you can use a book to help you identify it.

A **field guide** is a book that contains descriptions of organisms grouped in various ways. You may have seen or used a field guide to flowers, birds, trees, or seashells. A field guide to trees gives the scientific name of a tree and a description that can help you identify the tree. It may have a map that shows where the tree is found. It may also show what the leaf, flower, and twig of the tree look like.

Student identifying leaves by using field guides and keys

A **key** lists choices describing characteristics of organisms. Each grouping in a key describes organisms that share a characteristic they received from a common ancestor. For instance, in the key below, birds with webbed feet may have a webbed-footed ancestor in common.

TRY THIS **Activity!**

What Bird Is This?

Use the key given to identify the bird shown. Choose one statement from each pair below. Each statement directs you to another pair of statements, which leads you to identifying the organism.

What You Need

pencil, *Activity Log* page 27

Key to Birds

1. Webbed feet........Go To 3
   No webbed feet.....Go To 2

2. Hooked bill......... Red-tailed hawk- *Buteo jamaicensis*
   No hooked bill.......Cardinal- *Cardinalis cardinalis*

3. Flat bill.............Mallard duck- *Anas platyrhynchos*
   No flat bill.......... Go to 4

4. Pouch............ Brown pelican- *Pelicanus occidentalis*
   No pouch..........Red-faced cormorant- *Phalacrocorax urile*

Use the characteristics of cells to make a key in your ***Activity Log***. This key should identify the organism in a kingdom based on cell type. Use the characteristics of nucleus, chlorophyll, number of cells, and food.

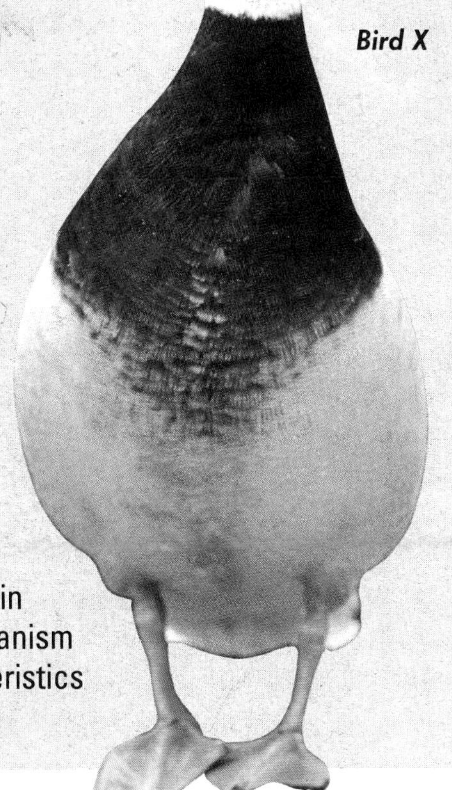

Bird X

Focus on Environment

What's Undiscovered?

There are probably more than 10 million living things in the world. With this in mind, can you believe that scientists have only classified about 1.5 million of them? Believe it or not, new organisms are being discovered every day. Because more than 50 percent of all plant and animal species in the world live in tropical rain forests, this is where many newly discovered organisms are found.

Unfortunately, the rain forests are being cut for farming or lumber. As the forest disappears, the habitats of these organisms disappear also. This affects not only the plants and animals that live there, but also those that migrate to the rain forests. The possible uses of these organisms in medicines, gasoline substitutes, cancer cures, and other purposes will never be known if they're destroyed before they're discovered. Scientists predict this loss will have serious consequences for planet Earth.

Rain forest, Costa Rica

Biology: Plants, Animals, and Ecology

While reading the book *Biology: Plants, Animals, and Ecology* by Ifor Evans, what organisms did you learn about? In your ***Activity Log*** on page 28, make five columns with the name of one of the five kingdoms at the top of each column. In the appropriate columns, list all the organisms from the book. How many of each kingdom did you find? How many different organisms did your class come up with?

Sum It Up

The only way we can study and understand all of the organisms on Earth is to classify them into groups. Because there are many evolutionary similarities between organisms, they can be grouped into a system of five kingdoms. After organisms are classified by kingdom, then they can be grouped by their shared evolutionary characteristics into one of their subgroups. People use these classifications to identify, organize, and show the evolutionary relationships of new organisms. Scientific naming allows us to communicate globally.

Garibaldi

Critical Thinking

1. What features or structures could people use to classify living organisms?
2. How do you think scientists would classify an organism that is newly discovered? Describe what they would do.
3. Why is classification important?

Theme T SCALE and STRUCTURE

WORKING TOGETHER TO SURVIVE

You are made of many cells that work together as tissues, organs, and organ systems. They carry out life functions in your body.

All the things we have learned about the structure of a cell, its parts, how cells divide, and how to classify living organisms into kingdoms has taken scientists many, many years to understand. The invention of the microscope during the 1600s gave scientists greater insight into living things they never knew existed before.

Looking through a microscope, you can see the smallest living things. You can see single cells made of many parts working together to carry on life processes.

Literature Link

A Time to Fly Free

In the book *A Time to Fly Free,* Josh Taylor spent many hours exploring the river near his home. He learned a great deal about the birds and other organisms that depended on the water for their survival. Make a list in your ***Activity Log*** on page 29 of all the organisms Josh knew about and observed near the river. What others can you think of that live in or near the water? Add them to your list. Compare your list with your classmates'. Then discuss how these organisms depend on each other for survival. Think about the shelter they need, the food they eat, and the environment they need to raise their young.

You may have noticed that your classmates' lists include different organisms than your list. But no matter what you named, you all had to consider the same basic needs, like shelter and food. All living things have these needs to survive, but each organism has a different way of obtaining them. How do you have needs like a bird? Both you and the bird have these basic survival requirements because you are both made of the same basic structures. The basic structures of all organisms are cells.

Great Blue Heron

Crow

Pied Kingfisher

Josh learned that some birds needed special help from people in order to survive. He worked with his friend Rafferty to save some birds' lives.

Seagull

Reddish Egret

The Basic Structures of Life

The cell is the basic unit of life. It takes millions of cells to form many of the living things you see in our world today. These cells combine to form tissues and organs in some organisms. Some of these tissues and organs work together as a system to provide nutrients so that an organism may live. Some of them form systems to carry out other basic life functions. Whether an organism is one-celled or many-celled, it performs life functions. Some organisms combine cells from two parents to reproduce and others divide to reproduce.

Each day scientists discover and classify new organisms in our world. All living things are made of cells that grow, develop, reproduce, and carry out life functions.

Minds On! Look carefully at this photograph. Now look back at pages 54–55. Can you tell what kingdom these cells are part of? ●

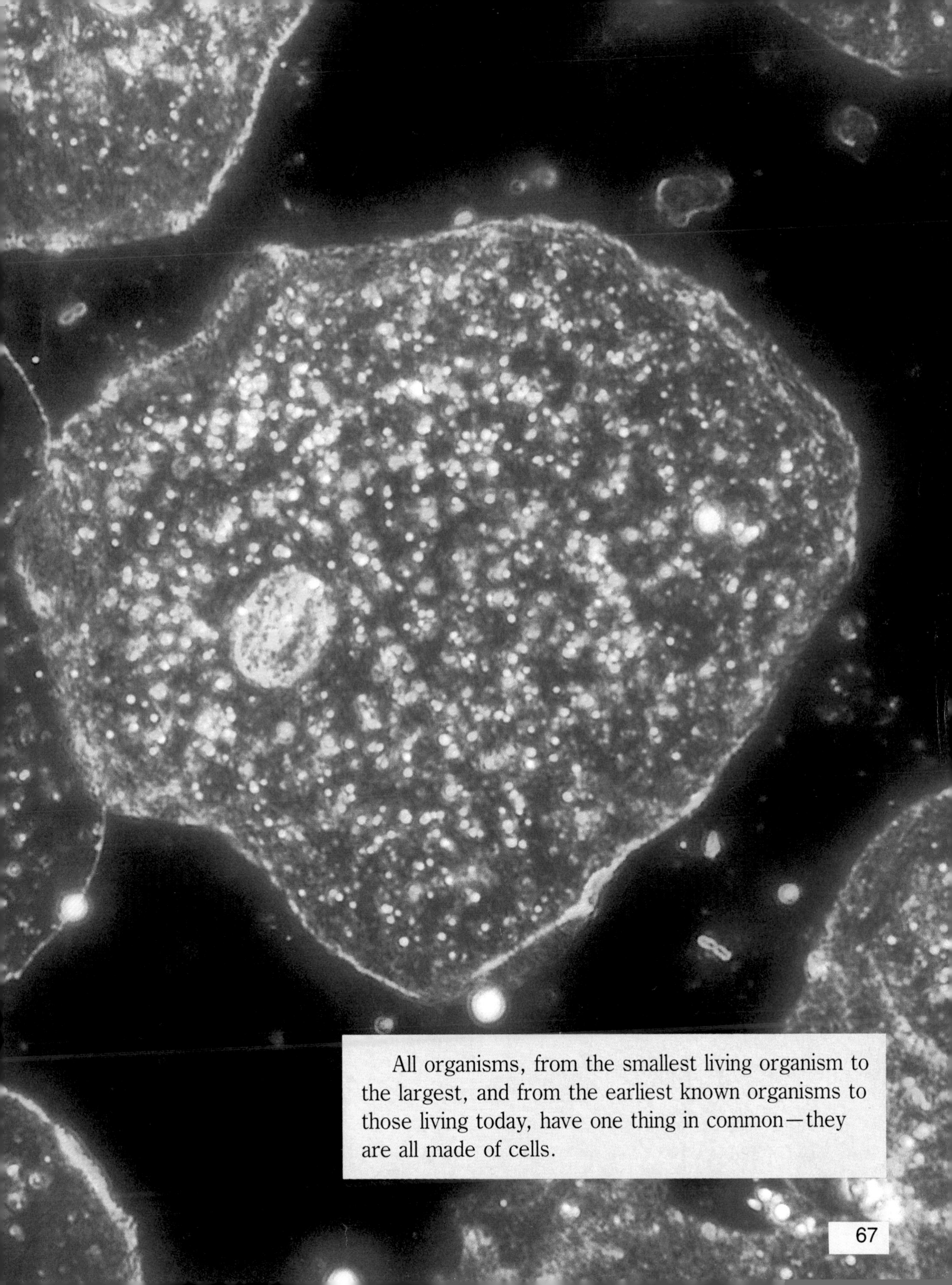

All organisms, from the smallest living organism to the largest, and from the earliest known organisms to those living today, have one thing in common—they are all made of cells.

GLOSSARY

Use the pronunciation key below to help you decode, or read, the pronunciations.

Pronunciation Key

a	at, bad
ā	ape, pain, day, break
ä	father, car, heart
âr	care, pair, bear, their, where
e	end, pet, said, heaven, friend
ē	equal, me, feet, team, piece, key
i	it, big, English, hymn
ī	ice, fine, lie, my
îr	ear, deer, here, pierce
o	odd, hot, watch
ō	old, oat, toe, low
ô	coffee, all, taught, law, fought
ôr	order, fork, horse, story, pour
oi	oil, toy
ou	out, now
u	up, mud, love, double
ū	use, mule, cue, feud, few
ü	rule, true, food
u̇	put, wood, should
ûr	burn, hurry, term, bird, word,courage
ə	about, taken, pencil, lemon, circus
b	bat, above, job
ch	chin, such, match
d	dear, soda, bad
f	five, defend, leaf, off, cough, elephant
g	game, ago, fog, egg
h	hat, ahead
hw	white, whether, which
j	joke, enjoy, gem, page, edge
k	kite, bakery, seek, tack, cat
l	lid, sailor, feel, ball, allow
m	man, family, dream
n	not, final, pan, knife
ng	long, singer, pink
p	pail, repair, soap, happy
r	ride, parent, wear, more, marry
s	sit, aside, pets, cent, pass
sh	shoe, washer, fish mission, nation
t	tag, pretend, fat, button, dressed
th	thin, panther, both
th̲	this, mother, smooth
v	very, favor, wave
w	wet, weather, reward
y	yes, onion
z	zoo, lazy, jazz, rose, dogs, houses
zh	vision, treasure, seizure

amputation (am′ pyə tā′ shən) the removal of a body part by surgery.

ancestor (an′ ses tər) an organism from which another organism is descended.

animal kingdom a group of many-celled organisms; some have structures to move from place to place.

asexual reproduction (ā sek′ shü əl rē′ prə duk′ shən) the production of offspring from only one parent.

cell smallest unit of living matter.

cell membrane (mem′ brān) a thin layer of proteins and fats that surround and contain cell parts.

cell wall a rigid layer that surrounds the cell membrane of plants, fungi and some bacteria.

chlorophyll (klôr′ ə fil′) a green pigment in plants that enables plants to make food by absorbing light energy.

chloroplast (klôr′ ə plast′) cell structure in which photosynthesis takes place; contains chlorophyll needed for photosynthesis.

chromosomes (krō′ mə sōmz′) cell parts inside the nucleus that carry the information that determines characteristics of an organism.

class a classification grouping of organisms smaller than a phylum.

classify to organize things into groups.

cytoplasm (sī′ te plaz′ əm) the jellylike liquid in a cell where chemical reactions of the cell take place.

egg sex cells formed by meiosis in females.

embryo (em′ brē ō′) a many-celled organism in its early stages of development.

evolution (ev′ ə lü′ shən) a process of change that happens to organisms over time.

family a classification grouping of organisms smaller than an order.

fertilization (fûr′ tə lə zā′ shən) the joining of a sperm and egg cell to form a zygote.

field guide a book that contains descriptions of organisms grouped in various ways for identification.

fungus kingdom (fung′ gəs king′ dəm) a group of many-celled or one-celled organisms that live in one place and obtain food from the materials on which they grow.

Hooke, Robert (1635–1703), an English scientist who first observed cells.

genus (jē′ nəs) a classification grouping of organisms smaller than a family.

key an identification tool for living and nonliving things made up of a list of paired statements based on certain characteristics.

kingdom (king′ dəm) the largest classification grouping of organisms.

Leeuwenhoek, Anton van (1632–1723) (lā′ vən hůk′) a Dutch scientist, one of the first to discover microscopic life; he recorded the first clear description of bacteria.

Linnaeus, Carolus (lə nē′ əs) (1707–1778) Swedish naturalist and botanist; developed the system of naming organisms using the genus and species name.

meiosis (mī ō′ sis) a type of cell division that produces cells and has half the number of chromosomes.

mitosis (mī tō′ sis) cell division in which one cell divides into two identical cells, each having the same number and type of chromosomes as the original cell.

moneran kingdom (mə nîr′ ən) a group of one-celled organisms that lack a nucleus and that either make their own food or obtain food from other organisms.

nucleus (nü′ klē əs) a dense, dark structure containing materials that control the activities of the cell.

order a classification grouping of organisms smaller than a class.

organ a group of tissues that work together to perform one or more life activities.

photosynthesis (fō′ tə sin′ thə sis) the process by which plants use light energy to produce food.

phylum (fī′ ləm) a classification grouping of organisms smaller than a kingdom.

plant kingdom a group of many-celled organisms that live in one place and make their own food through photosynthesis.

prosthesis (pros thē′ sis) an artificial replacement for a lost or amputated part of the body.

protist kingdom (prō′ tist) a group of one-celled organisms that make their own food or eat other organisms.

reproduction (rē′ prə duk′ shen) the process by which a living thing produces new organisms like itself.

scientific name the genus and species name of a living organism.

sexual reproduction (sek′ shü əl rē′ prə duk′ shən) the production of offspring by two parents.

species (spē′ shēz) the smallest and most specific classification grouping of an organism; its members are more closely related to each other than to any other organism.

sperm (spûrm) sex cells formed by meiosis in males.

tissue a group of similar cells that together perform a special job.

zygote (zī′ gōt) a single cell, produced by fertilization, that grows by cell division to become a complete many-celled organism.

INDEX

CREDITS

Photo Credits:

Cover, ©Manfred Kage/Peter Arnold, Inc.; **1,** ©Marty Snyderman; **2-3,** (b) Tom Walker/Stock Boston; **3,** (t) Janet Adams Photography/1991; **5,** ©Superstock; **6-7,** ©Tom Walker/Stock Boston; **8-9,** Telegraph Colour Library/FPG International; **10-11,** Superstock; **10-11,** (i) ©Studiohio; **14-17,** ©Platinum Studios; **19,** (tl) Ken Highfill/Photo Researchers, (tr) Lynwood Chase/Photo Researchers, (c) ©Stephanowicz/Panographics; (bl) E.R. Degginger, (br) Mark Wright/Photo Researchers; **20,** (cr) ©Martin Rotker/A.M. Siegelman, (br) ©Platinum Studios; **21,** (tr) ©Platinum Studios, (cr) Pasieka/Zefa/H. Armstrong Roberts, Inc., (bl) Comstock Inc.; **22,** (t) Art Wolfe/ALLSTOCK, (bl) Kenneth Garrett/FPG International, (br) K.S. Studios/1991; **23,** (bl) Biophoto/Photo Researchers, (tl) ©Eric V. Graves/Phototake; **24,** (tr) Chuck O'Rear/Westlight, (bl) Motion Control, Inc.; **25,** (tr) ©Frederick Bodin/Stock Boston, (br) ©Manfred Kage/Peter Arnold, Inc.; **26-27,** ©Dr. Mark Moffett; **28-29,** ©Janet Adams; **31,** ©Dwight Kuhn; **32,** (br) Bio. Associates/Photo Researchers (bl) Carolina Biological Supply Company; **33,** (bl) ©Studiohio, (bc) ©Stephanowicz/Panographics, (br) ©Mike Mazzaschi/Stock Boston; **34,** (tr) Earth Scenes/Dr. N. Smith, (bl) Marche/FPG International; **35,** (tl, tc, tr, bc) ©David Scharf/Peter Arnold, Inc., (br) ©Manfred Kage/Peter Arnold, Inc.; **36-37,** Mitch Reardon/Tony Stone Worldwide/Chicago Ltd.; **38-39,** ©Platinum Studios; **41,** (tl, tc, tr) ©Matt Meadows, (c) ©J. Albertson/Stock Boston, (b) Comstock Inc.; **42,** Carolina Biological Supply Company; **43,** (t) A.M. Siegelman/FPG International, (c) ©Platinum Studios; **45,** (t) ©A.M. Siegelman; **46,** ©Bob Daemmrich; **47,** (tr) ©H. Armstrong Roberts, Inc., (br) Superstock; **48,** ©BioSurface, Inc.; **49,** ©Bob Daemmrich; **50,** (tl) ©George Riley/Stock Boston, (bl) The Image Bank/Nicholas Foster; (bc) ©Dr. Tony Brain/Phototake, (br) L. West/FPG International; **51,** (tl)Earth Scenes/Patti Murray, (tc) Earth Scenes/Fred Whitehead, (tr) Tom McHugh/Photo Researchers, (bl) A.M. Siegelman; **52,** ©Studiohio; **53,** (tl) ©Studiohio, (tr) ©Ed Reschke/Peter Arnold, Inc., (cl) ©Studiohio, (c) Runk Schoenberger/Grant Hielman Photography, Inc.; (cr) ©Herbert Wagner/Phototake, (b) ©Studiohio; **54,** (bla) Comstock Inc., (blb) Frink/Waterhouse/H. Armstrong Roberts, Inc., (blc) ©J. Dunn/Stock Boston, (bld) Kevin Morris/ALLSTOCK (ble) R.A. Clevenger/Westlight, (blf) ©Renie Batory/The Stockhouse; **55,** (tl) Herbert Wagner/Phototake, (tlc) Runk/Schoenberger/Grant Heilman Photography, Inc., (cl) Dr. Tony Brain/Photo Researchers, (c) E.R. Degginger/H. Armstrong Roberts, Inc., (cr) ©Manfred Kage/Peter Arnold, Inc., (bl) Earth Scenes/Michael Fogden, (br) Earth Scenes/E.R. Degginger; **56,** Schmidt/FPG International; **58,** (tr) Zig Leszczynski/Animals Animals, (cr) The Image Bank/Wienberg/Clark, (cl) E.R. Degginger/H. Armstrong Roberts, Inc., (bl) Comstock Inc., (bc) E.R. Degginger/H. Armstrong Roberts, Inc., (br) Patti Murray/Animals Animals; **59,** ©Doug Martin; **60,** ©Platinum Studios; **61,** John Walden/ALLSTOCK; **62,** The Image Bank/James Carmichael; **63,** ©Marty Snyderman; **64,** ©Bob Daemmrich; **65,** (tc) A. Schmidecker/FPG International, (tr) Lee Kuhn/FPG International, (c) Stan Osolinski/FPG International, (bl) Frank Cezus/FPG International, (br) Lee Kuhn/FPG International, **66-67,** ©Jim Solliday/Biological Photo Service.

Illustration Credits:

12, 13, 57, Keith Freeman; **14, 28, 38, 52,** Bob Giuliani; **16, 18, 20, 30, 31, 32, 40(b), 44, 45,** Nancy Heim; **17, 50, 51, 54, 55,** Jan Wills; **40(l), 58, 59,** Anne Rhodes